Inconsistent Mathematics

Mathematics and Its Applications

Managing Editor:

M. HAZEWINKEL

Centre for Mathematics and Computer Science, Amsterdam, The Netherlands

Volume 312

Inconsistent Mathematics

by

Chris Mortensen
Centre for Logic,
Department of Philosophy,
University of Adelaide,
North Terrace, Australia

KLUWER ACADEMIC PUBLISHERS
DORDRECHT / BOSTON / LONDON

Library of Congress Cataloging-in-Publication Data

Mortensen, Chris.
    Inconsistent mathematics / Chris Mortensen.
        p.   cm. -- (Mathematics and its applications ; v. 312)
    Includes bibliographical references and index.

    1. Logic, Symbolic and mathematical. 2. Inconsistency (Logic)
I. Title. II. Series: Mathematics and its applications (Kluwer
Academic Publishers) ; v. 312.
QA9.M752 1995
510'.1--dc20   94-37996

ISBN 978-90-481-4480-8

Published by Kluwer Academic Publishers,
P.O. Box 17, 3300 AA Dordrecht, The Netherlands.

Kluwer Academic Publishers incorporates
the publishing programmes of
D. Reidel, Martinus Nijhoff, Dr W. Junk and MTP Press.

Sold and distributed in the U.S.A. and Canada
by Kluwer Academic Publishers,
101 Philip Drive, Norwell, MA 02061, U.S.A.

In all other countries, sold and distributed
by Kluwer Academic Publishers Group,
P.O. Box 322, 3300 AH Dordrecht, The Netherlands.

Printed on acid-free paper

All Rights Reserved
© 1995 Kluwer Academic Publishers
Softcover reprint of the hardcover 1st edition 1995
No part of the material protected by this copyright notice may be reproduced or
utilized in any form or by any means, electronic or mechanical,
including photocopying, recording or by any information storage and
retrieval system, without written permission from the copyright owner.

TABLE OF CONTENTS

ACKNOWLEDGEMENTS

Chapter Five substantially in the present form appears in *The Notre Dame Journal of Formal Logic*, (Vol.31 No.2), Spring 1990. Chapter Eleven sections 1-3 are with Peter Lavers. Chapter Twelve is by William James. Chapter Fourteen sections 3-5 are with Joshua Cole, and written by him. The research and publication were aided by a series of research grants from The University of Adelaide. The research was also aided by a Visiting Fellowship at the Automated Reasoning Project, Australian National University in August – September 1989; thanks to Michael McRobbie for making available the resources of the Automated Reasoning Project. I wish to thank Steve Leishman for his invaluable research assistance, William James for the LaTeX diagrams and Mary Mattaliano for the LaTeX typing. Thanks to Allen Hazen, as well as to Richard Benham, Edwin Coleman, Steve Leishman and other members of the University of Adelaide Logic Group, for useful comments and corrections on an earlier version. Finally I wish to thank Catherine Speck for her support and encouragement while this was being written.

The University of Adelaide

25.4.94

CHAPTER 1: MOTIVATIONS

1. Paraconsistency

The following idea has recently been gaining support: that the world is or might be inconsistent. In its modern manifestation it has been the province of rigorous symbolic logic, with motivations from logic, semantics and the foundations of mathematics. However, the idea finds roots in an older view, that change especially motion is contradictory, which can be traced back through Engels and Hegel to Zeno and Heraclitus, and has recently been revived, *e.g.* by Priest [46].

Two recent convergent motivations have been the paradoxes of logic, semantics and set theory, and the semantics of relevant logic. A theory is a set of sentences closed under a deductive relation (but cf. Definition 2.8). A logic (Def 2.7) is then a theory with the extra property of being closed under the rule of uniform substitution (relative to a specified set of connectives, such as 'and', 'or', 'not', 'if..then', 'for all', 'there exists', 'equals'). This expresses the idea that the logic of a collection of connectives ought to be neutral as to subject matter. The point stressed here is that deductive theories come with a logic in the background, albeit one which is perhaps tacitly presupposed as natural.

Let us consider set theory first. The most natural set theory to adopt is undoubtedly one which has unrestricted set abstraction (also known as naive comprehension). This is the natural principle which declares that to every property there is a unique set of things having the property. But, as Russell showed, this leads rapidly to the contradiction that the Russell set (the set of all non-self-membered sets) both is and is not a member of itself. The overwhelming majority of logicians took the view that this contradiction required a weakening of unrestricted abstraction in order to ensure a consistent set theory, which was in turn seen as necessary to provide a consistent foundation for mathematics. But all ensuing attempts at weakening set abstraction proved to be in various ways ad-hoc. Da Costa [10] and Routley [51] both suggested instead that the Russell set might be dealt with more naturally in an inconsistent but nontrivial set theory (where triviality means that every sentence is provable, see Definition 2.4). Since the triviality of a theory undoubtedly makes it uninteresting, this means that the background logic of any inconsistent theory should not validate the rule *ex contradictione quodlibet* (ECQ: from A and not-A to deduce any B), which two valued Boolean logic validates. Da Costa proposed C_1 as one such logic, and demonstrated some C_1-independence results for set theories containing the Russell set. Graham Priest argued persuasively that semantic paradoxes such as The Liar, and set-theoretic paradoxes such as Russell's, are best solved by accepting that there are some true contradictions; rather than sacrificing the generality and naturalness of principles like an unrestricted truth predicate or unrestricted set abstraction (*e.g.* [45],[46]). The need follows for a logic in which ECQ fails. Such logics are known as inconsistency-tolerant, or *paraconsistent*.

A second motivation came from Anderson and Belnap's investigations of *relevance* or conceptual connection. The idea was that correct natural entailment rests on conceptual connection; so that ECQ could not be a universally valid principle, because its premises can evidently be irrelevant to its conclusion (B may have no connection to A and not-A). However, subsequent discoveries in the semantics of sentential relevant logics by Belnap-Dunn-Fine-Meyer-Plumwood-Routley-Urquhart made it clear that the existence of inconsistent theories was necessary for relevance (though not sufficient). On this point the 1972 paper by Routley and Routley [52] was one of the earliest and most telling.

One can distinguish two strands of doctrine here: *strong paraconsistentism* is the acceptance of true contradictions, while *weak paraconsistentism* is the thesis that contradictory possibilities or structures have to be considered in the semantics of natural logic. Within strong paraconsistentism one can locate the two views we began with. First there is the modern motivation that true contradictions arise by *a priori* argument from various paradoxes; (for example, the argument that the Liar sentence 'This sentence is false' is demonstrably both true and false; or the argument that the Russell set is demonstrably both a member of itself and not a member of itself.) Second there is the older motivation that a philosophically correct account of change forces true contradictions on us. Within weak paraconsistentism, on the other hand, one can distinguish the thesis that inconsistent semantic structures represent genuine possibilities; so that while no contradictions are true, some contradictions are possible.

Two further distinctions are worth making here. First, we began by noting the thesis that the world is or might be inconsistent. But inconsistency is, strictly, a property of linguistic things like propositions or theories: the 'not-A' and the 'A' of a contradiction are the kinds of things capable of being true or false, that is propositions or sentences. So one can instead have a linguistic version of the central thesis: that the one true and exhaustive theory of the universe is inconsistent. The universe would be inconsistent, then, just to the extent that its true and exhaustive theory was inconsistent. We will not really need to make anything of this distinction in this book, though it can be said that the approach is certainly to study theories and other language-like items. A second distinction which has been made, is between viewing contradictions as propositions which are both true and false, and viewing them as true propositions of the form A and not-A (see e.g. Meyer and Martin [26]). Again, we will not be making much of this possible difference in this book. But it should be noted that to facilitate formal study, the mark of inconsistency in a theory is taken to be the presence among its consequences of the propositions A and not-A.

The attraction of the Russell set is that of providing a foundation of mathematics on a simple principle such as the naive principle of comprehension. An alternative contradictory foundation might be found in category theory, which looks interestingly close to inconsistency in places (see Chapter 11), as any broad abstraction principle will. Mathematicians undoubtedly flirt with such principles. There are, it should be noted, technical problems with inconsistent naive comprehension: while the Russell set can be tolerated, a stronger paradox, Curry's paradox, threatens triviality for deeper reasons (Meyer-Dunn-Routley [29], Slaney [53]). Brady has done important work demonstrating that the ordinal structure in such a set theory does not collapse in [5].

These ideas are foundationalist in spirit, while this book is not foundationalist. Many working mathematicians (though by no means all) are suspicious of logicians' apparent attempt to take over their subject by stressing its foundations. Surely one can reasonably feel that contradiction in set theory or category theory could not remotely threaten the immense corpus of mathematical results and applications discovered over more than two millenia. That alone is an argument for the naturalness of limiting the spread of contradictions. I do not mean that no foundationalist program for mathematics could succeed, however. A foundationalist program might succeed, and the inconsistent versions look the most promising. Indeed if the special case thesis defended later is true (that consistent complete mathematics is a special case of general mathematics), then any successful foundationalist program should be inconsistent or paraconsistent in some sense. But I have been persuaded by Edwin Coleman [8] that foundationalism in mathematics should be regarded with considerable suspicion; or at least that proper 'foundations', arguably both formalist and conventionalist in broad senses, would be much more complex and semiotical than twentieth century mathematical logic has attempted. In which case it would be arguable whether 'foundations' is an appropriate term.

The first consciously inconsistent number-theoretic structure seems to have been Meyer's inconsistent arithmetic modulo two (see below Chapter 2); though it is fair to say that his main concern was with demonstrating the consistency of the relevant arithmetic $R\#$, rather than with inconsistency for its own sake [23],[24],[25]. Routley replicated the result using a different nonclassical background logic [51]. Dunn's admirable paper [11] considered three-valued paraconsistent model theory with applications to arithmetic and type theory. Priest and Routley called for inconsistent infinitesimals [47] on the grounds that inconsistent claims abound in the pre-Weierstrassian history of the calculus. See also Asenjo [6]. It is argued here that without a properly developed inconsistent calculus based on infinitesimals, then inconsistent claims from the history of the calculus might well simply be symptoms of confusion. This is addressed in Chapter 5. It is further argued that mathematics has a certain primacy over logic, in that paraconsistent or relevant logics have to be based on inconsistent mathematics. If the latter turns out to be reasonably rich then paraconsistentism is vindicated; while if inconsistent mathematics has serious restrictions then the case for being interested in inconsistency-tolerant logics is weakened. (On such restrictions, see this chapter, section 3.) It must be conceded that fault-tolerant computer programming (*e.g.* Chapter 8) finds a substantial and important use for paraconsistent logics, albeit with an epistemological motivation (see this chapter, section 3). But even here it should be noted that if inconsistent mathematics turned out to be functionally impoverished then so would inconsistent databases.

2. Summary

In Chapter 2, Meyer's results on relevant arithmetic are set out, and his view that they have a bearing on Gödel's incompleteness theorems is discussed. Model theory for nonclassical logics is also set out so as to be able to show that the inconsistency of inconsistent theories can be controlled or limited, but in this book model theory is kept in the background as much as possible. This is then used to study the functional properties of various equational number theories.

Chapter 3 considers equational theories constructed from inconsistent models modulo an infinite prime. Chapter 4 introduces order. In the first section it is shown that the result of classical model theory that the theory of dense order without endpoints is $\aleph_0$-categorical, breaks down in the inconsistent case. In the second section arithmetical functions are added and results about ordered rings and fields are summarised.

In Chapter 5, a congruence relation on the ring of noninfinite hyperreal numbers is defined; leading to an inconsistent theory in which reasonable notions of limits, continuity, differentiation and integration can be defined, and Taylor's formula and polynomial differentiation proved. A simple change to this theory produces an incomplete differential calculus. In both theories every function is continuous. The latter theory is shown to have some similarity with synthetic differential geometry, a well-known incomplete theory in the language of categories due to Lawvere, Kock and others.

In Chapter 6 we begin by considering Priest's use of the Leibniz Continuity Condition (LCC) to produce an inconsistent theory of motion. The LCC is seen to have further implications. It is then shown that inconsistent functions are definable which can be regarded as everywhere continuous derivatives of certain classical functions which are not everywhere differentiable from the classical point of view. Chapter 7 puts together the previous two chapters to produce a structure in which functionality fails though in a controlled way; and uses this for an account of delta functions, which can in turn be regarded as inconsistent derivatives of the inconsistent continuous functions of Chapter 6.

Chapter 8 applies the additive group ideas of Chapter 6 to the solution of inconsistent systems of linear equations, and implications for control theory are outlined. Chapter 9 briefly considers the case of inconsistent vector spaces which suffer similar limitations to inconsistent fields. However inconsistent projective spaces over these vector fields do not suffer the same limitations. In inconsistent projective geometry modulo an infinite prime, it is shown that the usual projective duality theorem can be extended to a stronger language.

In Chapter 10, inconsistent quotient topologies are studied. It is shown that there is an interaction between classical topological concepts and the functionality of certain inconsistent topological spaces.

In Chapter 11, consistency problems for category theory are briefly surveyed. Then an important type of category, namely toposes, are studied. It is shown that a simple dualising operation gives rise to topos-like structures whose natural logic is not the usual intuitionist open-set logic, but rather its topological dual closed-set logic, which is paraconsistent. In Chapter 12, this open-closed duality is pursued further into the theories of presheaves and sheaves. In Chapter 13 extra dualising operations are considered, and it is shown that again inconsistent and incomplete theories can be regarded as duals. In the light of these dualities, it is argued that inconsistent and incomplete theories are deserving of equal respect.

Finally, in Chapter 14 we look briefly at the foundations as they are traditionally conceived. First, looking at the concept of provability, the fate of the Gödel sentence in the setting of inconsistent arithmetic is studied. Then, turning to the concept of the truth predicate, Kripke's nontriviality result using an incomplete theory is surveyed. It turns out that it is easily adapted to produce an inconsistent theory which represents the Liar sentence as a true inconsistency. Finally, turning to set theory containing unrestricted abstraction, we survey Brady's use of a fixed point method similar to Kripke's to produce an inconsistent set theory in which the Russell set is inconsistently self-membered. Here, the duality is seen to work in reverse, in that an incomplete set theory which does not decide on the self-membered status of the Russell set and other non-well-founded sets, can readily be constructed.

3. Philosophical Implications

To paraphrase Marx: philosophers have hitherto attempted to understand the nature of contradiction, the point however is to change it. Recent debates on the rule Disjunctive Syllogism (DS: from (A or B) and not-A to deduce B) have centered on whether according to the natural logic of mathematicians that rule is valid. In most of the structures of this book, DS fails. In view of the well known Lewis arguments, paraconsistentists are committed to denying that DS is valid since it leads quickly to the validity of ECQ. (Proof: take as premisses A and not-A. From A deduce (A or B) by the Principle of Addition. Then using (A or B) and not-A

deduce B by Disjunctive Syllogism.) However, giving up DS does not seem to be such a heavy burden, as argued in [31] and [32] (see also [6],[7],[20],[49]). In any case, if reasonable mathematics can be developed without DS and ECQ, then the claim that they are universally logically valid is weakened.

The question of the validity of DS and ECQ has not always been distinguished from the question of whether mathematicians are *habitual consistentisers*. Now it seems to me that mathematicians are indeed habitual consistentisers, at least in regarding inconsistency as implying something wrong with the premises. However, this needs some qualification: mathematical practice is not absolutely univocal on the matter. For example, consider the history of infinitesimals in pre-Weierstrass calculus, or the old quantum theory (Bohr theory of the atom), or delta functions both before and after Schwartz, or the terminology of 'identification' in quotient constructions (see later chapters). As well, as noted earlier, there are persistent tendencies among mathematicians to use very general abstraction principles for both sets and categories, which can lead to inconsistency in short order. And of course there is the semantical thinking about self reference which has mostly been the province of logicians. But it is argued here that consistency can be relaxed without complete disorder resulting. In turn this casts doubt on any attempt to argue back from habitual consistency to the logical validity of DS and ECQ.

These claims apply to inconsistent mathematics considered as pure mathematics. There are, I suggest, at least three justifications for studying inconsistent mathematical theories. The first justification might be called the *argument from pure mathematics*. The argument from pure mathematics for studying inconsistency is the best of reasons: because it is there. In other words, nothing in this book relies on the thesis that contradictions are true. Nor is it claimed that the mathematics needed to describe existing physical systems is inconsistent. But then, how could you be perfectly sure? Just possibly a physical reason might be found, or perhaps some pleasing metaphysical reason (for example, the thesis that inconsistent calculus gives a better theory of motion). It is always dangerous to think that a physical use will *never* be found for a given piece of mathematics. Nor is present-day mathematical physics anomaly-free: witness the singularities at the beginning of time or in black holes, delta functions in elementary quantum theory, or renormalisation in quantum field theory.

These observations amount to a second reason for studying inconsistent theories, what can be called the *ontological justification*. This is, essentially, the paraconsistentist claim that a contradiction is true or might be true, backed up by one's favourite arguments from semantics or physics. *Ontology* here means having to do with how things *are*; and contrasts with *epistemology*, which has to do with how things are *known*.

Thus there is also a third reason for studying inconsistent theories, what can be called the *epistemological justification*. This is the argument that any *information system* with more than one source of information must permit the possibility of conflict between its sources. Here it can be conceded that the world is consistent, so that an inconsistent database would inevitably be incorrect somehow. But it is not always easy to produce a consistent cut-down of one's information base, at least one which is not arbitrary in its selection of what to discard. Yet humans display the ability to operate in an anomalous data environment. This is plainly because evolving creatures face real-time difficulties; it might just take too long to solve the problem of what is the truth about one's environment. Evidently, informationally-sensitive machines face similar problems. For example, an aircraft aloft might be receiving contradictory data from its sensors but be unable to take the luxury of shutting down before solving the problem of what is the nature of the physical environment. Thus, rules for operating deductively in an inconsistent data environment are necessary, and the rule ECQ which permits the deduction of everything from an inconsistency is unhelpful. At the very least, a study of inconsistent theories in which such a rule is broken is indicated as part of the long-term goal of artificial intelligence. Thus, even if the world is consistent, and having to deal with inconsistencies turns out to be because of the epistemic limitations of finite humans or computers, inconsistency-tolerance might well remain a permanent part of a good expert system.

A special case thesis was mentioned earlier. No claim is being made that inconsistent or incomplete mathematics is better than classical mathematics. Still, classical mathematics is a special case in two ways. To the extent that the logic of classical mathematics is classical two-valued logic, then classical logic is a special case of paraconsistent logic, holding over a restricted domain in which the assumption of the truth-preservingness of DS holds. Further, general mathematics can be consistent or inconsistent, complete or incomplete, prime or nonprime. (A theory is nonprime when it contains some disjunction without containing the disjuncts: Meyer's $R\#$ interestingly turned out to be nonprime, and there is nonprime quantum arithmetic, on both of which see Chapter 2; as well, classical Peano arithmetic is nonprime by Gödel's theorem.) Nontrivial classical theories satisfy the special assumption of DS, so nothing is lost because it is all there in the classical special case. The aim of the present work is to expand conceptions of mathematics, not to deny the obviously excellent corpus of classical mathematics.

Intuitionist mathematics is the home of incomplete theories (but note that any consistent axiomatisable classical theory containing arithmetic is also incomplete by Gödel's theorem). The present point of view is firmly in favour of intuitionist and/or incomplete mathematics (see especially Chapters 5, 6 and 11). Let a thousand theories bloom. However the main concern in this book is with inconsistency, if for no other reason than that incompleteness has seemed easier to swallow than inconsistency, something not so easy to justify given the duality results of Chapters 11-13. Those results turn on the topological duality between open sets and closed sets. It is well-known that intuitionist logic is also the logic of open sets; that is, that intuitionist logic stands to open sets as classical two valued Boolean logic stands to sets in general. It is less well-known that the logic of closed sets is paraconsistent, and this is considered especially in Chapter 11. Intuitionist theories have sometimes been very complicated. The three-valued approach to incomplete theories (below,

Chapter 5) can often illustrate many of their features in a simple and natural way. But conversely the complex brilliancy of incomplete theories in, say, the theory of sheaves, or synthetic differential geometry, is highly admirable. A three-valued incomplete approach to the truth predicate was taken by Kripke in his excellent paper 'Outline of a Theory of Truth' [19], which is discussed in Chapter 14. Kripke's approach has been nicely generalised to the inconsistent incomplete case by Fitting [12].

It can also be said that the use of closed-set logics for various theories below contributes to the re-vindication of the point of view of Brazilian logic. Brazilian logic studied paraconsistent logics partly with the aim of dualising intuitionism (see Da Costa [10]). For a time, however, it looked as if these logics suffered a serious limitation, because they do not admit a reasonable implication operator (see [30] and [36]). However, closed-set logic can fairly be described as Brazilian-style. Also, a somewhat cavalier attitude is taken to implication in later chapters: to the extent that implication is the converse of deducibility, the latter is usually preferred below. Mathematics, unlike logic, seems not very interested in nested implications, and even less interested in nesting of depth three or more (perhaps it should be). This is in line with the position defended later in this section and elsewhere, that mathematics is functional rather than logical. And needless to say, existing theories based on Brazilian logics (see above first section) are as legitimate and interesting as any other.

I would further argue that the only way to establish validity of the paraconsistent point of view is to demonstrate the existence of a rich and interesting inconsistent mathematics. Without that, the paraconsistent position would seem to rest on a motivation at best epistemic and computational, and functionally impoverished at that. That is why finding a distinctive inconsistent perspective on analysis, the crown jewel of classical mathematics, is desirable. The present book falls somewhat short of that; again for functional reasons the real numbers look to be essentially consistent (see especially Chapters 2-5 below). It is to be hoped that the situation with inconsistent analysis will ultimately improve. Combining inconsistency with incompleteness would seem to be the right way to go here. For an example of this combination, see Chapter 6.

Fortunately or unfortunately, the methods and results in this book indicate that the 'essence' of mathematics is deeper than paraconsistentists have thought, though it is also argued here that this is no vindication of any classical consistent limitation. Definitions are necessary to explain this point. A theory is *functional* iff, if an equation $t_1 = t_2$ holds then Ft_1 holds iff Ft_2 holds, where F is any atomic context (roughly, F is any 'logic-free' context lacking $\&, \vee, \sim, \rightarrow, \forall, \exists$); and a theory is *transparent* iff the same is true except that F is any context, including possibly the logical operations as well. (See Definition 2.11.) Now in the following chapters it is clear that being at least functional, if not transparent, is a good constraint to have on a theory. Without it, control over identity seems to be lost and one wonders what equality stands for, especially in the equational subtheory (though in a few theories later there is a controlled relaxation of functionality, see *e.g.* Chapter 7). But a method of proving inconsistent theories to be functional is commonly to find existing consistent theories which are invariably functional, add appropriate denials of atomic sentences using a paraconsistent background logic, and let the functionality of the latter ride in on the back of the consistent theory. Indeed it is not hard to show that any inconsistent complete functional theory has a consistent complete functional subtheory, which one might expect millenia of classical mathematics to have encountered. So one might say that classical mathematics, interested in functionality, concentrated on the consistent functional subtheory, naturally failing to notice related inconsistent supertheories.

At its strongest this might be the criticism that inconsistent mathematics leads to no new functional insights, a typical mathematician's complaint. But I do not want to concede too much here: the situation is nowhere near as bad as that. For one thing, one can say that inconsistency is functionally no worse than consistency. The consistency constraint is unnecessary and binding, and full functionality is available without it. One might further argue that this shows that functionality is deeper than consistency, completeness, or primeness. The question of the validity of DS and ECQ is irrelevant to the essence of mathematics, one might say. This also gives the prospect of searches for partly functional structures, where the failure of functionality is controlled by a combination of incompleteness and inconsistency. But also, it is not true that there are no interactions between functionality and inconsistency or incompleteness. The down side of this is the difficulty of inconsistentising real number theory, the essential consistency of the real and hyperreal number fields which was just mentioned. But the up side is that this can lead to interesting insights about functionality; for example Chapter 10 on topology, and any other time a quotient construction is done by thinking of 'identifying' distinct elements (Chapters 9,10 and elsewhere).

The essentialist talk above about the essence of mathematics should not be taken too literally. It is only intended to claim that abandoning functionality is a bigger departure from mathematics as it is practised, than abandoning consistency or completeness is. If anything, inconsistent mathematics suggests antiessentialism and antinecessitarianism about mathematical truth. By necessitarianism is meant here the claim that there is a special unshakeable status for the truths of mathematics and logic, that true mathematics and logic *cannot* have been false. I think the argument here against necessitarianism proceeds by attempting to whiteant the necessitarian's resolve, rather than knockdown refutation. But if consistency is not a necessary constraint on rigorous mathematical reasoning, then surely nothing is sacred. For further argument in favour of antinecessitarianism, also called possibilism, see [35] or [38].

In place of necessitarianism, it seems right to put conventionalism about mathematical truth. By this is meant the idea that mathematics, particularly pure mathematics, is more like a decision than a discovery of a pre-existing truth. I do not mean the kind of conventionalism that confers a mystical power on conventions or decisions to inaugurate or sustain necessary or logical truth. The old game analogy does seem the right one: the 'truths' of pure mathematics are internal to mathematical theories in a way like rules are internal to games, and quite unlike the way empirical claims about the physical world are true or false.

There are of course some interesting disanalogies: mathematics seeks unity between its branches, while games seem quite happy to be isolated from one another. Also, certain mathematical propositions have a preferred status (accepted) over others (denied), whereas preferred strategies can be reversed in different games. But whatever the explanation here, it presumably lies in the social. The question wherein lies rigor presumably has an explanation in terms of mathematical society and its relations to the rest of society and to the physical world and its properties (forms). Certainly an explanation in terms of the necessity of certain propositions, as opposed to their mere truth, would add nothing.

Another conclusion to draw, perhaps more speculatively, is antiplatonism about the abstract objects of pure mathematics. Of course, the relationships between the propositions of mathematics and the forms (properties) of objects in the physical world, are very complex. Still there seems to be no difficulty in principle in denying timelessly existing abstract mathematical objects as truthmakers. One can be gripped by the picture of eternal objects, if one thinks of theories as necessarily consistent. Freeing up consistency suggests ultimate freedom from any constraint. But the absence of external constraints is the mark of fiction, not of fixed existing truthmakers. Convention cannot create what is not there to begin with.

CHAPTER 2: ARITHMETIC

1. Relevant Arithmetic

The first consciously inconsistent arithmetical structure can fairly be dated at Robert K. Meyer's 1975 nontriviality proof for the consistent relevant arithmetic $R\#$. In deference to history and to whet the reader's appetite, this chapter begins with a simplified version of his argument. We will see that it has considerable significance for the understanding of Gödel's incompleteness theorems. Later in the chapter, the style of argument used by Meyer will be extended so as to study a broader range of number systems. First we need a formal language. This is the standard type of first order language used for studying formalised arithmetic theories, save that for technical reasons a distinction is made between two implication operators $\supset$ and $\to$. The language has names for all the natural numbers, $0, 0', 0'', \ldots$, the arithmetical operations $+$ and $\times$, and the usual logical apparatus.

Definition 2.1. The *language* $\mathcal{L}$ contains a single binary relation symbol $=$, a single constant 0, term-forming operators $+, \times, '$, (the latter is read 'the successor of'). $0^{(n)}$ is defined as $0'^{\cdots'}$ with n superscripted dashes. That is, $0^{(n)}$ is a name for the natural number n. In addition, there is logical apparatus consisting of variables $x, y, z, \ldots$, a unary propositional operator $\sim$, binary propositional operators $\&, \vee, \supset, \equiv, \to, \leftrightarrow$, and quantifiers $\forall$ and $\exists$. The number of primitive operators can be reduced by defining $A \vee B \ =df \ \sim (\sim A \& \sim B)$, $A \equiv B \ =df \ (A \supset B)\&(B \supset A)$, $A \leftrightarrow B \ =df \ (A \to B)\&(B \to A)$, and $(\exists \ ) \ =df \ \sim(\forall \ )\sim$. The connective $\to$ is not material implication $\supset$. The latter can be defined by $A \supset B \ =df \ \sim A \vee B$; the former is intentional and not definable from $\&, \vee$, and $\sim$. 0 is a term, and if t_1 and t_2 are terms, so are t_1', $(t_1 + t_2)$, and $(t_1 \times t_2)$; and $t_1 = t_2$ is an atomic sentence. All atomic sentences are sentences, and if A and B are sentences, so are $\sim A$, $A\&B$, $A \vee B$, $A \supset B$, $A \equiv B$, $A \to B$, $A \leftrightarrow B$, $(\forall x)Ax$, and $(\exists x)Ax$, where x is a variable not occurring in A and Ax is the result of replacing some term in A wherever it occurs by x.

We study theories of various logics. Theories are deductively closed sets of sentences (of the above language in the present case). Different deduction relations are thus possible, which are described by the axioms and rules of different logics. Meyer used RQ, a quantified version of the well-known relevant logic R.

Definition 2.2. The *relevant logic* RQ is given by all universal closures of the following axiom schemata and rules.

Axioms:

(1) $(A \rightarrow B) \rightarrow ((B \rightarrow C) \rightarrow (A \rightarrow C))$

(2) $A \rightarrow ((A \rightarrow B) \rightarrow B)$

(3) $(A\&B) \rightarrow A$

(4) $(A\&B) \rightarrow B$

(5) $((A \rightarrow B)\&(A \rightarrow C)) \rightarrow (A \rightarrow (B\&C))$

(6) $A \rightarrow (A \vee B)$

(7) $B \rightarrow (A \vee B)$

(8) $((A \rightarrow C)\&(B \rightarrow C)) \rightarrow ((A \vee B) \rightarrow C)$

(9) $(A\&(B \vee C)) \rightarrow ((A\&B) \vee (A\&C))$

(10) $\sim\sim A \rightarrow A$

(11) $(A \rightarrow\sim A) \rightarrow\sim A$

(12) $(\forall x)Ax \rightarrow At$ (t any term)

(13) $(\forall x)(A \rightarrow B) \rightarrow ((\forall x)A \rightarrow (\forall x)B)$

(14) $A \rightarrow (\forall x)A$ (x not free in A)

(15) $(\forall x)(A \vee B) \rightarrow (A \vee (\forall x)B)$ (x not free in A)

(16) $((\forall x)A\&(\forall x)B) \rightarrow (\forall x)(A\&B)$

Rules:

(17) If A and B are theorems so is $A\&B$

(18) If A and $A \rightarrow B$ are theorems so is B.

To obtain the logic RMQ add the axiom scheme $A \rightarrow (A \rightarrow A)$.

The logic R (the sentential fragment of RQ) is an important relevant logic. A logic is said to be *relevant* iff, whenever $A \rightarrow B$ is a theorem of the logic then A and

B share an atomic sentence in common. Relevant logics were studied extensively by Anderson and Belnap in [1] as well as by many others; and were proposed as serious rivals to classical (two-valued) logic as a logic of natural reasoning, which is arguably relevant at least in its sentential fragment. Every valid argument of R is a valid argument of classical logic, but not vice-versa. With a language and a logic, we can now specify arithmetical theories, in this case Meyer's relevant arithmetic $R\#$. $R\#$ is similar to classical Peano arithmetic, called here $P\#$, save that in several places material implication $\supset$ is replaced by the strong implication $\to$ used by all the usual relevant logics.

Definition 2.3. The *arithmetic $R\#$* is given by:

logical axioms and rules are those of RQ,

arithmetical axioms:

(#1) $(\forall x, y)(x = y \leftrightarrow x' = y')$

(#2) $(\forall x, y, z)(x = y \to (x = z \to y = z))$

(#3) $(\forall x)(\sim x' = 0)$

(#4) $(\forall x)(x + 0 = x)$

(#5) $(\forall x, y)(x + y' = (x + y)')$

(#6) $(\forall x)(x \times 0 = 0)$

(#7) $(\forall x, y)(x \times y' = (x \times y) + x)$

Arithmetical rule of mathematical induction (RMI): If $F0$ and $(\forall x)(Fx \to Fx')$ are theorems, so is $(\forall x)Fx$.

For *the arithmetic $RM\#$*, add the logical axiom scheme $A \to (A \to A)$. For the *arithmetics $R\#\#$ and $RM\#\#$*, add to $R\#$ and $RM\#$ respectively Hilbert's rule Ω: if $F0, F1, F2\ldots$ are all theorems, so is $(\forall x)Fx$. For *classical Peano arithmetic $P\#$*, take as logical axioms and rules those of classical quantification theory, and replace $\to$ and $\leftrightarrow$ in (#1), (#2) and (RMI) by $\supset$ and $\equiv$ respectively. For *classical standard arithmetic $P\#\#$*, add rule Ω to $P\#$.

To the extent that R captures a more plausible account of implication than classical logic does, as argued by Anderson, Belnap and many others, then it is arguable that $R\#$ gives a closer account of 'natural' arithmetic than classical Peano arithmetic $P\#$ does. This does not 'deny' $P\#$: it is an interesting system like many another, not the least because the Gödel incompleteness theorems can be proved of it (see below).

Definition 2.4. An arithmetical theory is (*negation*) *consistent* if for no sentence A is both A and $\sim A$ provable, else (*negation*) *inconsistent*; and *nontrivial* if not every sentence is provable, else *trivial*.

Proposition 2.5. (Meyer) $R\#$ is nontrivial.

Proof. The argument uses the three-valued logic $RM3$ with values $\{F, B, T\}$, the two element domain $\{0, 1\}$ of arithmetic modulo two, and an interpretation function I assigning terms to elements of $\{0, 1\}$ and sentences to elements of $RM3$. Set $I(0^{(n)}) = n \bmod 2$, $I(+) = +\bmod 2$, $I(\times) = \times \bmod 2$. For any terms t_1, t_2, set $I(t_1 + t_2) = I(+)(I(t_1), I(t_2))$, $I(t_1 \times t_2) = I(\times)(I(t_1), I(t_2))$, and $I(t_1') = (1 + I(t_1)) \bmod 2$. Set $I(t_1 = t_2) = B$ if $I(t_1) = I(t_2)$, else $I = F$. For nonatomic sentences, set $I(\sim A) = T$, B or F as $I(A) = F$, B or T respectively. Order $\{F, B, T\}$ by $F < B < T$; and then set $I(A\&B) = \min\{I(A), I(B)\}$; set $I(A \vee B) = \max\{I(A), I(B)\}$; set $I(A \to B) = I(\sim A \vee B)$ if $I(A) \le I(B)$, else $I(A \to B) = I(\sim A\&B)$; and set $I((\forall x)Fx) = \min\{y : \text{for some term } t, I(Ft) = y\}$. It is now a straightforward, if lengthy, argument to verify that all theorems of $R\#$ take one of the values $\{B, T\}$. But since $0 \bmod 2 \ne 1 \bmod 2$, $I(0 = 1) = F$. Hence $R\#$ is nontrivial. $\square$

Meyer makes the point that this argument relies only on methods which are finitistic in Hilbert's sense. In particular, the quantifiers can be treated by establishing that $I((\forall x)Fx) = I(F0\&F1)$, which is a standard argument in the metalanguage. Thus $R\#$ enjoys an advantage over classical Peano arithmetic $P\#$: that its nontriviality can be established by finitistic methods. Yet as Meyer pointed out, all primitive recursive functions are representable in $R\#$, which is thus subject to the Gödel incompleteness theorems also. But this is not really a puzzle. The explanation is that relevant and other paraconsistent logics turn on making a distinction between negation inconsistency and triviality, the former being weaker than the latter; whereas classical logic cannot make this distinction. For what the present author's intuitions are worth, these do seem to be different concepts. Thus for $R\#$, negation consistency cannot be proved by finitistic means by Gödel's second theorem, whereas nontriviality can be shown. Since $P\#$ collapses this distinction, both kinds of consistency are infected by the same unprovability.

Indeed $R\#$ can do even better; for consider any 'false' equation $t_1 = t_2$, *i.e.* one which reduces by calculation to $0^{(n)} = 0^{(m)}$ where these are classically distinct natural numbers. Then a simple modification of the above proof using modulo$(\max\{m, n\} + 1)$ instead of modulo 2, shows that $t_1 = t_2$ is not a theorem of $R\#$. $R\#$ has thus greater security of calculation than $P\#$. If there is negation inconsistency in $R\#$, it is along way away, contained. Meyer uses these conclusions to call for a revived Hilbert program based on relevant rather than classical logic: since there exist finitistic proofs that various undesirable conclusions do not follow in relevant arithmetic, mathematics based on relevant logic can escape the limitations of Gödel. The Hilbertian point of view is not, however, taken in the present work.

One might also wonder whether there is the prospect of the nontriviality of $P\#$, and hence its negation consistency, via a proof that $P\# \subset R\#$. Needless to say such an argument would be nonfinitistic by Gödel 2, but it might be interesting nonetheless. This is the *gamma problem for* $R\#$. It turns out that if $R\#$ were closed w.r.t. the rule: if A and $A \supset B$ are theorems so is B, then $P\# \subset R\#$. Recently however Meyer has shown that this rule fails for $R\#$. Meyer has expressed some dissatisfaction with $R\#$ because of this result, but it seems that it makes $R\#$ all the more interesting.

This is not a book about relevance, however. The point of view might be described as paraconsistent, insofar as that implies an interest in nontrivial inconsistent theories.

2. Nonclassical Logics and Their Theories

We proceed to a more systematic approach to inconsistent theories. In order to show that inconsistency can be contained in a deductively rigorous way, it is necessary to set out some of the theory of models. In this section we define theories and the semantical notion of an assignment. In the next section we bring in the notion of a domain and thus a model. Models are best thought of as devices for *controlling the membership of theories*; only secondly are they what theories are *about*. Theories define their own 'aboutness', regardless of domains of interpretation. An analogy is with the values of a many-valued logic, which are in the first place devices for controlling membership of theories. We begin by extending the language of the previous section, so as to deal with a broader range of applications.

Definition 2.6. The *language* $\mathcal{L}$ has the following components:

(1)   a collection of *names* or *atomic terms*, for example names for some or all the rational numbers, or integers, or rational numbers, or real numbers, or hyperreal numbers or other mathematical entities. Normally these names are taken from the entities themselves, *i.e.* self-naming. For each theory the collection of names must be specified.

(2)   a collection of n-ary *term-forming operators*, for example from among $+, \times, -, /, ^{-1}, ', (\ ,\ ,\ )$, and possibly others. If $t_1, \ldots, t_n$ are terms and $\circ$ is an n-ary term forming operator, then $\circ t_1, \ldots, t_n$ is a term. Term-forming operators are also called *function symbols*.

(3)   a collection of n-ary *primitive predicates*, for example from among $=, <, \varepsilon$ and possibly others. If $t_1, \ldots, t_n$ are terms and F is an n-ary primitive predicate then $Ft_1 \ldots t_n$ is an *atomic sentence*.

(4)   a collection of n-ary *sentential operators*. The general purpose negation symbol is $\sim$, but $\neg$ (open set intuitionist negation) and $\ulcorner$ (closed paraconsistent negation) are sometimes used; also sentential operators $\&, \vee, \supset, \equiv, \rightarrow, \leftrightarrow$.

(5)   *variables* x, y, z, and possibly others; and *quantifiers* ($\forall$) also written ( ),
      and ($\exists$). These form nonatomic or complex sentences in the usual way. It is
      stipulated that no term is a variable and that only theories containing just
      closed sentences are considered.

Definition 2.7. A *logic L* is a set of sentences closed under uniform substitution and under a consequence relation $\vdash_L$, sometimes written $\models_L$ if the logic is
semantically specified. The subscript is dropped if it is clear what logic is intended.

Definition 2.8. An *L-semitheory* (of logic *L*) is a set Th of closed sentences
satisfying: if $A \in Th$ and $A \vdash_L B$ then $B \in Th$. An *L-theory* is an *L*-semitheory
also satisfying: if $A \in Th$ and $B \in Th$ then $A\&B \in Th$. If Th is an *L*-theory
and $A \in Th$ then we write $\vdash_{Th} A$, dropping the Th when no confusion will result.
An *L*-semitheory Th is $\sim$-*inconsistent* if for some A both $A \in Th$ and $\sim A \in Th$
(similarly for $\neg$ and $\ulcorner$, inconsistent for short); else Th is $\sim$-*consistent* (consistent
for short). Th is *prime* if, whenever a disjunction $A \vee B$ is in Th at least one of
the disjuncts is too. Th is *incomplete* if for some sentence A neither A nor $\sim A$
is in Th, else Th is *complete*; and Th is *trivial* if $Th = \mathcal{L}$, else *nontrivial*. Th is
zero-degree or *extensional* if none of its members contains occurrences of $\rightarrow$ or $\leftrightarrow$.
Most of the theories in this book are extensional. If $Th_1 \subseteq Th_2$, we say that Th_2
extends or *is an extension of* Th_1.

Logics and theories in this book are determined by specifying (1) a lattice,
complete in the lattice-theoretic sense, (2) a definition of the operator $\sim$ (or $\neg$ or
$\ulcorner$), and a definition of the relation $\vdash$ or the operator $\rightarrow$ or both, on the lattice,
(3) a subset ∇ of designated values, closed upward under the order on the lattice,
(4) an assignment function I : language $\mathcal{L} \rightarrow$ the lattice.

In more detail:

Definition 2.9. An *RM3-theory* is determined by:

(1) The lattice whose Hasse diagram is

$$
\begin{array}{c}
\circ\ \ \mathrm{T} \\
|\ \\
\circ\ \ \mathrm{B} \\
|\ \\
\circ\ \ \mathrm{F}
\end{array}
$$

That is, $A \cup B = \mathrm{lub}\{A, B\}$ and $A \cap B = \mathrm{glb}\{A, B\}$.

(2) For $\rightarrow$ and $\sim$,

$\rightarrow$	T	B	F	$\sim$
T	T	F	F	F
B	T	B	F	B
F	T	T	T	T

(3) Set of designated values $\nabla = \{B, T\}$.

(4) $I : \mathcal{L} \rightarrow \{F, B, T\}$ is a partial function satisfying

   (4.1) If A is an atomic sentence and $I(A)$ is defined, then $I(A) \in \{F, B, T\}$

   (4.2) $I(A\&B) = I(A) \cap I(B)$

   (4.3) $I(A \vee B) = I(A) \cup I(B)$

   (4.4) $I(\sim A) = \sim I(A)$

   (4.5) $I(A \rightarrow B) = I(A) \rightarrow I(B)$

   (4.6) $I((\forall x)Fx) = \mathrm{glb}\{y : \text{for some term } t, I(Ft) = y\}$

   (4.7) $I((\exists x)Fx) = \mathrm{lub}\{\text{the same set}\}$

Such a function I is called an *RM3-assignment*. A semitheory Th is then determined by the condition $Th =df \{A : I(A) \in \nabla\}$; and if ∇ is a filter on $RM3$ then Th is a theory. One says that I *determines* Th. If Th is the $RM3$-theory determined by I, then A *holds in* Th and I if $I(A) \in \nabla$. The *logic RM3* is the set of sentences which hold in all such I (*e.g.* all instances of $A \rightarrow A$), together with the definition: $A \vdash B =df$ for all I, if $I(A) \in \nabla$ then $I(B) \in \nabla$.

For *P3-theories*, make only two changes (a) $B \rightarrow B = T$, (b) $\sim B = T$. This is often signalled by changing $\sim$ to $\ulcorner$. Such theories are complete and generally inconsistent and nontrivial. The *logic P3* is defined in the same way as *RM3* above.

For *J3-theories*, (a) change the letter '*B*' in *P3* to '*N*' (for 'neither'), (b) change $\sim$ to $\neg$, (c) $\neg N = F$, (d) $\nabla = \{T\}$. These theories are consistent and generally incomplete. The *logic J3* is generated as above.

For *PJ4-theories*, use the lattice

Also, (a) $\sim T =\sim N = F$, $\sim F =\sim B = T$ ($\sim$ is used for negation here since it is 'neutral', but note that *P3* and *J3* are sublogics), (b) $A \rightarrow B =df\ T$ if $A \le B$ else $A \rightarrow B = F$, (c) $\nabla = \{B, T\}$. This generates theories which are in general both inconsistent and incomplete. The *logic PJ4* is defined as above.

For the next few chapters *RM3*-theories are mostly used. While theories of other logics are interesting and have different properties from *RM3*-theories, many of the interesting functional questions are invariant w.r.t. changes between the logics. *P3* and *J3* theories are used in the chapter on differential calculus. *PJ4*-theories are used in the chapters on inconsistent continuous functions and on linear equations.

A simple result is the following:

Proposition 2.10. (Extendability lemma) Let I_1 and I_2 be *RM3*-assignments in the same language, and Th_1 and Th_2 be the extensional theories generated. If the atomic sentences holding in I_1 are a subset of those holding in I_2, and the negations of atomic sentences holding in I_1 are a subset of those holding in I_2, then Th_2 is an extension of Th_1, i.e. $Th_1 \subseteq Th_2$.

Proof. By induction on the number of occurrences of $\{\sim, \&, \forall\}$. Note first that the hypothesis of the proposition is equivalent to the following: If A is any atomic sentence then   (a) if $I_1(A) = T$ then $I_2(A) \in \{B, T\}$, (b) if $I_1(A) = B$ then $I_2(A) = B$, (c) if $I_1(A) = F$ then $I_2(A) \in \{F, B\}$. The inductive argument shows that (a)-(c) hold of all sentences. (Base clause:)   Already defined to be true. ($\sim$clause:)(a) If $I_1(\sim A) = T$ then $I_1(A) = F$, so by inductive hypothesis (c), $I_2(A) \in \{F, B\}$; whence $I_2(\sim A) \in \{B, T\}$. The cases (b) and (c) are similar. (&clause:) (a) If $I_1(A\&B) = T$ then $I_1(A) = I_1(B) = T$; whence $I_2(A\&B) \in \{B, T\}$. The cases (b) and (c) are similar. ($\forall$clause:) (a) If $I_1((\forall x)Fx) = T$ then for all terms t, $I_1(Ft) = T$; so by inductive hypothesis (a) for all terms t, $I_2(Ft) \in \{B, T\}$; whence $I_2((\forall x)Fx) \in \{B, T\}$. The cases (b) and (c) are similar. $\square$

Note that the proposition fails for theories containing $\rightarrow$. The result is applied in many places in what follows. A frequent strategy is to take a consistent complete classical extensional (zero-degree) theory and extend it by adding extra atomic sentences which were assigned F in the old theory. This amounts to choosing a new $RM3$-assignment in which the extra atomic sentences and their negations are all assigned B. The Extendability lemma then ensures that the new theory loses none of the sentences of the old.

Definition 2.11. A theory is *functional* iff, for all terms t_1, t_2, if $t_1 = t_2$ holds then for any atomic sentence Ft_1 containing t_1, Ft_1 holds iff Ft_2 holds, where Ft_2 is like Ft_1 except for replacing t_1 in one or more places by t_2. A theory is *transparent* if the same condition holds except that Ft_1 can be any sentence (not restricted to atomic).

Now notice that the Extendability lemma does not ensure that an inconsistent extension of a functional theory must itself be functional, and similarly for transparency. One cannot add classically false sentences to a theory willy-nilly and expect to remain functional. For example, if $0 = 2$ is added to classical natural number theory without also adding $0+1 = 2+1$, *i.e.* $1 = 3$, then functionality fails. And even if functionality were ensured, if $\sim 0 = 0$ is not also added then transparency fails (since $\vdash 0 = 2$ and $\vdash \sim 0 = 2$ but not $\vdash \sim 0 = 0$). Functionality and transparency coincide for classical consistent complete theories, but not for $RM3$,

$P3$, *etc.* It is a desirable characteristic for mathematical theories to be at least functional if not transparent, for then equality means something so to speak. In some later chapters it is seen that transparency is not as important as functionality, but also that even the latter can be relaxed in a controlled and reasonably-motivated way. From now on for several chapters, unless otherwise stipulated, we deal with $RM3$-assignments and their associated theories.

Definition 2.12. An assignment is *prereflexive* iff for all terms t_1 there is some term t_2 such that $t_1 = t_2$ holds; and *reflexive* iff for all terms t, $t = t$ holds. An assignment is *normal* iff (i) reflexive, and (ii) $t_1 = t_2$ holds iff $t_2 = t_1$ holds, and (iii) if $t_1 = t_2$ and $t_2 = t_3$ hold so does $t_1 = t_3$.

Proposition 2.13. (1) An assignment I is transparent iff for all terms t_1, t_2, if $t_1 = t_2$ holds then for all atomic F, $I(Ft_1) = I(Ft_2)$.
(2) If I is functional and prereflexive then I is normal.
(3) If I is prereflexive and Th is an equational theory, then I is transparent iff I is functional and for t_1, t_2, if $\vdash t_1 = t_2$ then for all t, $I(t_1 = t) = I(t_2 = t)$.

Proof. (1) R to L is a straightforward induction on the number of occurrences of $\{\sim, \&, \forall\}$ in sentences. L to R : Suppose for some t_1, t_2, and atomic F, $I(Ft_1) \neq I(Ft_2)$. If one of Ft_1 and Ft_2 does not hold then the other does so that I is not functional and thus not transparent. Otherwise, if both Ft_1 and Ft_2 hold then one of $\sim Ft_1$ and $\sim Ft_2$ does not hold and the other does, and again I is not transparent.

(2) Let t be any term. By prereflexivity, there is a term t_1 such that $\vdash t = t_1$. By functionality, $t = t$ holds iff $t = t_1$ holds, so $\vdash t = t$. Also by functionality, $t_1 = t_2$ holds iff $t_2 = t_2$ holds and $t_2 = t_1$ holds iff $t_2 = t_2$ holds. Hence if $t_1 = t_2$ holds then $t_2 = t_1$ holds. Finally, again by functionality, if $t_1 = t_2$ holds then $t_1 = t_3$ holds iff $t_2 = t_3$ holds; so if $t_1 = t_2$ and $t_2 = t_3$ hold then $t_1 = t_3$ holds. Hence I is normal.

(3) Let I be prereflexive and Th an equational theory. If I is transparent then certainly I is functional, and by (1) if $\vdash t_1 = t_2$ then for all t, $I(t_1 = t) = I(t_2 = t)$. Conversely, from (1) it suffices to prove that if $\vdash t_1 = t_2$ then for all atomic F,

$I(Ft_1) = I(Ft_2)$. Clearly it suffices to prove for just one replacement of t_1 by t_2 since an obvious induction then proves it for more replacements. Thus Ft_1 has one of two forms, $t(t_1) = t$ or $t = t(t_1)$, and Ft_2 has the corresponding forms $t(t_2) = t$ or $t = t(t_2)$. That is it has to be proved that $I(t(t_1) = t) = I(t(t_2) = t)$ and $I(t = t(t_1)) = I(t = t(t_2))$. Now by functionality, prereflexivity and (2), $\vdash t(t_1) = t(t_1)$; hence if $\vdash t_1 = t_2$ then by functionality $\vdash t(t_1) = t(t_2)$. Hence by the condition of the theorem, $I(t(t_1) = t) = I(t(t_2) = t)$. The other case is similar. $\square$

A simple result is the following

Proposition 2.14. (Term elimination) Let Th be a transparent extensional theory determined by I. For any terms t_1, t_2 in $\mathcal{L}$, let $t_1 \approx t_2$ iff $t_1 = t_2$ holds in Th or t_1 and t_2 are the same term, and let $\mathcal{L}'$ be any sublanguage of $\mathcal{L}$ containing just one term from each $\approx$-equivalence class and agreeing with $\mathcal{L}$ on function symbols and primitive predicates. Let Th' be the theory determined by the assignment I' which is the restriction of I to $\mathcal{L}'$. Then the restriction of Th to $\mathcal{L}' = Th'$.

Proof. By induction on the number of occurrences of $\{\sim, \&, \forall\}$.
(Base:) The atomic sentences of Th in the weaker language have the exact values they have in I'.
($\sim$ and & clauses:) Straightforward.
($\forall$ clause:) ($\forall 1.1$:) If $I((\forall x)Fx) = T$ then $I(Ft) = T$ for all t in $\mathcal{L}$. Hence $I(Ft) = T$ for all t in $\mathcal{L}'$, so $I'(Ft) = T$ for every t in $\mathcal{L}$, so $I'((\forall x)Fx) = T$.
($\forall 1.2$:) Conversely, if $I'((\forall x)Fx) = T$ then $I'(Ft) = T$ for every t in $\mathcal{L}'$. Since Th' is transparent, for every t in $\mathcal{L}$, $I(Ft) = T$; so $I((\forall x)Fx) = T$.
($\forall 2.1$:) If $I((\forall x)Fx) = B$, then $I(Ft) \in \{B, T\}$ for every t in $\mathcal{L}$, and for some t_1, $I(Ft_1) = B$. Hence $I(Ft) = I'(Ft) \in \{B, T\}$ for every t in $\mathcal{L}'$. But also $t_1 \approx t_2$ for some t_2 in $\mathcal{L}'$, and $I'(Ft_2) = I(Ft_2) = I(Ft_1) = B$; so $I'((\forall x)Fx) = B$.
($\forall 2.2$:) If $I'((\forall x)Fx) = B$, then for all t in $\mathcal{L}'$, $I'(Ft) \in \{B, T\}$; and for some t_1 in $\mathcal{L}'$, $I(Ft_1) = B$. But for all t_2 in $\mathcal{L}$ there is some t_3 in $\mathcal{L}'$ such that $t_2 \approx t_3$ and $I(Ft_2) = I(Ft_3) = I'(Ft_3) \in \{B, T\}$ with also $I(Ft_1) = B$; so that $I((\forall x)Fx) = B$.
The cases ($\forall 3.1$) and ($\forall 3.2$), when $I((\forall x)Fx) = F$, are similar. $\square$

Note that the last proposition fails if 'transparent' is weakened to 'functional': suppose that while $t_1 = t_2$ holds, for some atomic F $I(Ft_1) = T$ but $I(Ft_2) = B$. Then the choice of t_1 rather than t_2 for $\mathcal{L}'$ affects whether $(\exists x) \sim Fx$ holds in I' and Th', but the restriction of Th to $\mathcal{L}'$ is unaffected.

3. Models for Number Systems with Arithmetical Operations

We proceed to the notion of a model, and then study various arithmetical operations.

Definition 2.15. A *model* is a pair $\langle D, I \rangle$ where D is a set and I is an assignment which also (1) assigns to every name a member of D, and is onto D so that every member of D is named; (2) I assigns to every n-ary function symbol an n-ary partial function on D; (3) the assignment to complex terms is given by $I(ft_1 \ldots t_n) = I(f)(I(t_1) \ldots I(t_n))$; (4) I satisfies: $t_1 = t_2$ holds iff $I(t_1)$ and $I(t_2)$ are defined and equal. These have the effects that I is normal and functional. A model $\langle D, I \rangle$ is *transparent* iff I is transparent, *inconsistent (incomplete)* iff the associated theory is inconsistent (incomplete). A model is an *extension* of another if the associated theory of the first is an extension of that of the second; and *infinite* iff D has an infinite cardinal, else *finite*.

For example, consider the class of finite transparent models described in the first section of this chapter, in which all (extensional) sentences of the classical standard model for the $\{+, \times\}$ arithmetic of the natural numbers hold (see [27]). Take names for all the natural numbers $\{0, 1, 2, \ldots\}$ (as usual we can let these be the natural numbers themselves); function symbols are $\{+, \times\}$. The domain is the integers modulo m, *i.e.* $\{0, 1, \ldots, m-1\}$. For every name t, set $I(t) = t \bmod m$; and set $I(+) = + \bmod m$ and $I(\times) = \times \bmod m$. This determines $I(t)$ for every term t. Finally set $I(t_1 = t_2) = B$ if $I(t_1) = I(t_2)$, else $I(t_1 = t_2) = F$. In [27] these are called $RM3^m$ and it is proved that they determine theories which are inconsistent, complete, nontrivial, ω-inconsistent, ω-complete and decidable. It is also well known that the classical integers mod m permit a definition of additive inverse, minus n, as $(-n) \bmod m =df \ m - n \bmod m$ if $n \bmod m \neq 0$ and 0 otherwise; and thus subtraction as $n_1(-\bmod m)n_2 =df \ n_1(+\bmod m)((-n_2)\bmod m)$. So adding to the model names for all negative integers and setting $I(-) = -\bmod m$ determines $I(t) \bmod m$ for all integer terms t in the $\{=, -, \times\}$ language. If as before $I(t_1 = t_2) = B$ if $I(t_1) = I(t_2)$ else $I(t_1 = t_2) = F$, then the Extendability lemma can be applied to conclude that every sentence of the classical consistent complete (zero-degree) theory of the ring Z of integers holds. Also clearly the conditions of Proposition 2.13 are satisfied so the model is transparent.

Summarising,

Proposition 2.16. There are finite inconsistent complete transparent models in which every sentence of the classical consistent complete theory of the ring of integers Z holds. □

It is worth noting that in these models it is not in general true that if A holds and $A \supset B$ holds then B holds. In particular, $(A\& \sim A)\&((A\& \sim A) \supset B)$ might hold while B does not, *e.g.* if A is $0 = 0$ and B is $0 = 1$. Thus, as is characteristic of inconsistent nontrivial theories of paraconsistent logics, the rule DS in the form (if A holds and $\sim A \vee B$ holds then B holds) fails in these models. But this does not imply any loss of *classical* information; since clearly for any A and B, if $A\&(A \supset B)$ holds, and also holds back in the classical standard model of the integers, then B holds (because by a well known argument DS holds for any consistent complete theory).

An application of the term elimination lemma (Proposition 2.14) is that since these inconsistent models for Z are transparent, their simple terms can be cut down to $\{0, 1, \ldots, m - 1\}$; and then the assignment $I(t_1 = t_2) = B$ iff $I(t_1) = I(t_2)$ else $I = F$, has exactly the same sentences true as the assignment with names for all integers, in their common language.

Moving to division and the theory of fields, it is well known that finite classical arithmetic modulo a prime p allows a definition of the unique multiplicative inverse n^{-1} of every number $n \neq 0$ in $\{0, 1, \ldots, p-1\}$, and thus division via $n_1/n_2 =_{df} n_1(\times \bmod m)(n_2^{-1})$. These can be described by the classical transparent assignment with $I(/) = /\bmod p$, and $I(t_1 = t_2) = T$ if $I(t_1) = I(t_2)$ else $I = F$; and all the classical postulates and consequences of the theory of fields hold. For example, the following postulates, which axiomatise the classical theory of fields, all hold (see [50], p.130).

(1) $(x, y, z)(x + (y + z) = (x + y) + z)$

(2) $(x, y)(x + y = y + x)$

(3) $(x)(x + 0 = x)$

(4) $(x)(x - x = 0)$

(5) $(x, y, z)(x \times (y \times z) = (x \times y) \times z)$

(6) $(x, y)(x \times y = y \times x)$

(7) $(x)(x \times 1 = x)$

(8) $(x)(\sim x = 0 \supset x \times x^{-1} = 1)$

(9) $(x, y, z)(x \times (y + z) = (x \times y) + (x \times z))$

(10) $\sim 0 = 1$

Changing to $I(t_1 = t_2) = B$ if $I(t_1) = I(t_2)$ and applying the Extendability lemma gives:

Proposition 2.17. There are finite inconsistent transparent models in which every sentence of the classical theory of fields holds. $\square$

This cannot be strengthened to the conclusion that all the theory of the field Q of rationals with names for all of them holds: not every integer has an inverse defined, those with $I(t) = 0$ do not whereas all but 0 itself do in classical Q. This suggests a general difficulty in inconsistently extending fields, which proves to be the case. In the next chapter it is seen that this can be achieved in infinite fields modulo an infinite prime.

These structures give a solution to the following problem, raised by Graham

Priest. One wants postulates such as the Cancellation Law ([4], p.2):

$$(x)(\sim x = 0 \supset (y,z)(x \times y = x \times z \supset y = z))$$

to hold when moving from the classical theory of rings to the classical theory of integral domains and fields. But inconsistent fields such as the above have both $\sim 0 = 0$ and $(y,z)(0 \times y = 0 \times z = 0)$ holding. Yet one does not want to detach every $y = z$ or the theory is uninteresting. But one does want to detach $y = z$ for those x classically not identical with 0 (in Q or R say). However, in the inconsistent fields mod p, while both $\sim 0 = 0$ and $(y,z)(0 \times y = 0 \times z = 0)$ holds, still $y = z$ cannot be detached (*e.g.* obviously not $\vdash 0 = 1$). This is another symptom of the general undetachability of $\supset$. But also, if $t = 0$ does not hold classically, then $(y,z)(t \times y = t \times z \supset y = z)$ holds; and then if $t \times t_1 = t \times t_2$ holds classically, $t_1 = t_2$ can be detached (all by the Extendability lemma).

A useful general result proved by Dunn in [11] is as follows.

Proposition 2.18. (Dunn) Let A be an algebra $\langle D, o_1, \ldots, o_n \rangle$ where the o_i are operations on D. Let A' be a subalgebra of A and h be a congruence from A to A' with $h(x) = x$ for all x in A'. Then the classical equational theory Th_1 of A with names for all elements of D can be extended to an inconsistent transparent theory Th_2 using the assignment $I(t) = h(t)$ for all names t, and $I(t_1 = t_2) = B$ if $I(t_1) = I(t_2)$ else $I(t_1 = t_2) = F$.

Proof. That Th_2 is an extension of Th_1 follows by the Extendability lemma from the fact that if $t_1 = t_2$ holds classically then clearly $I(t_1) = I(t_2)$, so that $I(t_1 = t_2) = I(\sim t_1 = t_2) = B$. But also condition (3) for transparency in Proposition 2.13 is evidently satisfied, so Th_2 is transparent.   □

This can be applied to the $\{+, \times, /\}$ congruence from the non-negative rationals into itself given by $h(x) = 1$ if (classically) $x \neq 0$ and $h(0) = 0$, to give an inconsistent transparent nontrivial extension of the classical $\{+, \times, /\}$ theory of the non-negative rationals with names for all of them. But the attempt to bring in the negative rationals and subtraction while retaining all classical laws and functionality, wrecks the theory: there needs to be an additive inverse $-n$ for each n; but if classically distinct elements n_1, n_2 are identified inconsistently, then by functionality they should have the same additive inverse, so that $n_1 - n_2 = n_1 - n_1$ holds. If the usual classical laws also are to hold then $n_1 - n_1 = 0$ also holds. But division by 0 is prohibited, so division by $n_1 - n_2$ must similarly be undefined by functionality again. But division by $n_1 - n_2$ is permitted in the classical theory. The interaction between subtraction and division is the problem here, and it strengthens the suspicion that it is not so easy to inconsistentise fields.

4. Summary of Further Results in Arithmetic

These are proved in [27] or [41].

Proposition 2.19. $RM3^m$ can be axiomatised by: $RM\#$, plus $\vdash 0 = m$ plus $\vdash 0 = t \leftrightarrow 0 = 1$ for every t in $\{0, 1, \ldots, m-1\}$.

Proposition 2.20. In $R\#$, $\vdash (0 = n_1 \circ 0 = n_2) \leftrightarrow 0 = gcd(n_1, n_2)$, where $A \circ B =df \sim (A \rightarrow \sim B)$ and $gcd(n_1, n_2)$ is the greatest common divisor of n_1 and n_2.

Definition 2.21. $RM(2n+1)^m$ is the result of replacing $RM3$ as background logic in $RM3^m$ by the logic $RM2n + 1$ (see [27]) $RM3\omega = df \bigcap_{all\ m} RM3^m$. $RM\omega =df \bigcap_{all\ m,n} RM(2n+1)^m$.

Proposition 2.22. $RM\omega$ is inconsistent, incomplete, nontrivial and ω-inconsistent. Its extensional part is complete, and identical with the zero degree part of $RM3\omega$. Any inconsistent zero-degree extension of the zero-degree part of $R\#$ is complete. Not all inconsistent extensions of $R\#$ are extensions of $RM\#$. The nontheorems of $RM\omega$ are recursively enumerable. Problem: is $RM\omega$ decidable?

Definition 2.23. LRQ is the logic axiomatised by dropping the distribution axiom (9) from RQ (Definition 2.2). $LR\#$ is then formed by adding the Peano postulates (#1)-(#7) and rule RMI (Definition 2.3) to LRQ.

Proposition 2.24. Distribution is not provable in $LR\#$, but is provable in any inconsistent extension of $LR\#$. Problem: is every extensional instance of Distribution provable in $LR\#$?

CHAPTER 3: MODULO INFINITY

1. The Classical Denumerable Nonstandard Model of Natural Number Arithmetic

The classical consistent complete denumerable model of the natural numbers, $\{0, 1, 2, \ldots\}$, of order type ω, is also called the standard model of classical Peano arithmetic. This contrasts with the (classical consistent complete denumerable) *nonstandard* model. As is well known the latter has a domain of order type $\omega + \eta(\omega^* + \omega)$, consisting of an initial block isomorphic to $\{0, 1, 2, \ldots\}$ (called the finite natural numbers), with succeeding blocks of numbers (called the infinite natural numbers) isomorphic to the integers (order type $\omega^* + \omega$), the blocks themselves being densely ordered (order type η). Both models verify exactly the sentences of classical standard arithmetic $P\#\#$ (see Definition 2.3) in their common language. In this chapter we consider consistent and inconsistent theories which arise from the nonstandard model.

Given a finite (natural) number m, any infinite number can be uniquely represented as the sum of a multiple (possibly infinite) of m plus a unique natural number between 0 and $m - 1$. This gives a natural definition of modulo m for all infinite numbers. Hence the $\{+, -, \times\}$ modulo models of Chapter 2 can have added to them names for all nonstandard infinite numbers.

Classical nonstandard numbers are constructed in such a way that first order properties of the standard natural numbers continue to hold. One such property is that the modulus of any number can be taken w.r.t. any nonzero number:

$$(\forall m, x)(\sim m = 0 \supset (\exists y, z)(x = zm + y \& 0 \leq y \leq m - 1)).$$

That is,

$$(\forall m, x)(\sim m = 0 \supset (\exists y, z)(x = zm + y \& (\exists w)(w + y + 1 = m))).$$

Hence in the nonstandard model, for any infinite numbers m and x, there is a multiple of m no more than $m - 1$ below x. This evidently allows a consistent complete model modulo infinite m with functions $\{+, \times\}$, as well as additive inverses and subtraction. Also there are plainly inconsistent versions given by $I(t_1 = t_2) = B$ if $t_1 \bmod m = t_2 \bmod m$ else $I = F$, and by the Extendability lemma all of the classical consistent complete $\{+, -, \times\}$ theory of the integers Z continues to hold in the inconsistent models.

An interesting subclass of these structures arises from the facts that infinite numbers can have infinite divisors and that infinite primes with no divisors save themselves and unity exist. (Proof: Add to classical Peano arithmetic $P\#$ the axioms $(x)((\exists y)(x + y = p) \supset (x = 1 \vee x = p \vee \sim (\exists z)(x \times z = p))$, i.e. p is prime, and $\sim p = 1, \sim p = 2, \sim p = 3, \ldots$. Every finite subset of axioms has a classical model so by the compactness theorem the whole theory has a model. But p is not one of the finite numbers.) Now noting that 'x divides y' is definable as $(\exists z)(x \times z = y)$, we have that the well known theorem of standard number theory that

$$(x, y, p)((\text{prime } p \And p \text{ divides } x \times y) \supset (p \text{ divides } x \vee p \text{ divides } y))$$

holds also in the (classical) nonstandard numbers, and thus for infinite primes p. (See [4], p.19, Theorem 9.) But this is equivalent to saying that if $x \times y = 0 \bmod p$, then $x = 0 \bmod p$ or $y = 0 \bmod p$. It follows easily that classical modulo infinite p is an integral domain and obeys the Cancellation Law $(x, y, z)(\sim x = 0 \supset (x \times y = x \times z \supset y = z))$. (See [4], p.5, Theorem 1.) The usual argument ([4], p.41) can then be applied to show that every $x \neq 0$ in mod p has a unique multiplicative inverse $x^{-1} \bmod p$. (Proof: The products $x \times 0, x \times 1, \ldots, x \times (p - 1)$ are all in $\{0, \ldots, p - 1\}$ since $(\times \bmod p)$ is an operation; but they are all distinct by the Cancellation Law. Exactly one of them must be 1 therefore, and $x^{-1} = df$ the unique y such that $x \times y = 1 \bmod p$.) Further by the Cancellation Law, if two numbers have the same inverse then they are identical. Classical mod infinite p is thus a field, and $x(/ \bmod p)y = df \ x(\times \bmod p)(y^{-1} \bmod p)$. Its 'natural' order type is evidently $\omega + \eta(\omega^* + \omega) + \omega^*$, with a last member $p - 1$ and a final block of order type ω^*. That is, the final block is $\{\ldots, p - 2, p - 1\}$.

Some properties of consistent mod p are:

(1) A number is finite iff its additive inverse is in the final block.

(2) $(p-1)/2$ and $((p-1)/2)+1$ are additive inverses in the same block (Benham), as are $[(p-3)/2$ and $((p-3)/2)+3)]$, ..., and $[(p-(2n+1))/2$ and $((p-(2n+1))/2)+(2n+1)]$, and $[(p+(2n+1))/2$ and $(p+(2n+1)/2)-(2n+1)]$ etc.

(3) The multiplicative inverse of any finite number save zero is defined and infinite.

(4) 2^{-1} is $(p+1)/2$, and $(p-2)^{-1} = (-2)^{-1} = (p-1)/2$, which are in the same block and additive inverses.

2. Inconsistency

Now we move to inconsistency. Take names for all members of the classical nonstandard model (naming themselves). If t is a name, then set $I(t) = t \bmod p$, set $I(+) = +\bmod p$ and similarly for $\{-, \times, /\}$. Note that $I(t_1/t_2)$ is not defined if $I(t_2^{-1}) = 0$; but this never happens if t_2 is a nonzero finite number. The terms $2^{-1}, 3^{-1}, \ldots$ can be used as names for the reciprocals of the natural numbers, and the terms $t_1 \times t_2^{-1}$ for all finite names t_1, t_2 can be used as names for all the rational numbers. The assignment $I(t_1 = t_2) = B$ if $I(t_1) = I(t_2)$ else $I(t_1 = t_2) = F$, determines an inconsistent theory. By the Extendability lemma this is an extension of the classical theory of the integers; and also of the integers modulo p so that all the sentences of the classical theory of fields hold. Also it is not difficult to show that it is transparent. Hence we have a strengthening of the results of the previous chapter (see also [28] and [34]):

Proposition 3.1. There exist infinite inconsistent transparent models modulo an infinite prime in which hold all sentences of the classical consistent complete theory of the field of rationals Q, with names for all the rationals. □

In these theories $\vdash \sim 0 = 0$, since $I(0 = 0) = B = I(\sim 0 = 0)$. A different construction of $\{+, \times\}$ arithmetic enables that to be avoided, with inconsistency being confined to the infinite part of the diagram, and has some consequences for Fermat's Last Theorem. Choose an arbitrary infinite number m and for domain D take $D = \{0,1,2,\ldots\} \cup \{x : m \leq x \leq 2m-1\}$. If t is a name, set $I(t) = t$ if t is finite else set $I(t) = t \bmod m + m$. (In other words shift $t \bmod m$ along by m to get $I(t)$.) Set $I(t_1 + t_2) = t_1 + t_2$ if both are finite, else $I(t_1 + t_2) = ((t_1 + t_2) \bmod m) + m$. Set $I(t_1 \times t_2) = t_1 \times t_2$ if both are finite, else $I(t_1 \times t_2) = ((t_1 \times t_2) \bmod m) + m$. Set $I(t_1 = t_2) = T$ if $I(t_1) = I(t_2)$ and both are finite, set $I(t_1 = t_2) = B$ if $I(t_1) = I(t_2)$ and both are infinite, else set $I(t_1 = t_2) = F$. Call this model and the associated theory NSN.

Proposition 3.2. Every sentence holding in the classical standard and nonstandard models of the natural numbers holds in NSN. NSN is transparent.

Proof. If A holds in standard arithmetic then it holds in the classical nonstandard model. An inductive argument on the complexity of the term t_1 shows that if $t_1 = t_2$ holds in the nonstandard model then $I(t_1) = I(t_2)$.

(Base:) Let t_1 be any name. If t_1 is finite then if $t_1 = t_2$ holds then $I(t_1) = t_1 = t_2 = I(t_2)$. If t_1 is infinite then $I(t_1) = t_1 \bmod m + m = t_2 \bmod m + m = I(t_2)$.

(+clause:) If t_1 is $t_3 + t_4$, then if both t_3 and t_4 are finite then $I(t_3 + t_4) = t_3 + t_4 = t_1 = t_2 = I(t_2)$. Else, if one of t_3, t_4 is infinite then also t_2 which $= t_3 + t_4$ is infinite. Hence $I(t_3 + t_4) = m + (t_3 + t_4) \bmod m = m + t_2 \bmod m = I(t_2)$.

($\times$clause:) Similar. But if $I(t_1) = I(t_2)$ then $t_1 = t_2$ holds in NSN, so NSN is an extension of the classical standard and nonstandard models.

For transparency, by Proposition 2.13(1) it suffices to prove that if $t_1 = t_2$ holds in NSN then for all atomic F, $I(Ft_1) = I(Ft_2)$. Now $t_1 = t_2$ holds iff $I(t_1) = I(t_2)$, so it must be proved that if $I(t_1) = I(t_2)$ then for all atomic F, $I(Ft_1) = I(Ft_2)$. Again it suffices to prove this for just one replacement of t_1 by t_2, so let Ft_1 be $t(t_1) = t_3$ and Ft_2 be $t(t_2) = t_3$ (the case where Ft_1 is $t_3 = t(t_1)$ is similar). Now note that if $I(t_1) = I(t_2)$ then $t_1 \bmod m = t_2 \bmod m$. (Reason: if t_1 is finite then certainly $t_1 = t_2$; while if t_1 is infinite then $I(t_1) = m + t_1 \bmod m = I(t_2) = m + t_2 \bmod m$, so that since the addition is ordinary addition of nonstandard numbers, $t_1 \bmod m = t_2 \bmod m$.) But now the functionality of the classical modulus construction ensures that $t(t_1) \bmod m = t(t_2) \bmod m$ and so that $m + t(t_1) \bmod m = m + t(t_2) \bmod m$. If $t(t_1)$ is finite then $I(t(t_1)) = t(t_1) = t(t_1) \bmod m = t(t_2) \bmod m = t(t_2) = I(t(t_2))$, and if $t(t_1)$ is infinite then $I(t(t_1)) = m + t(t_1) \bmod m = m + t(t_2) \bmod m = I(t(t_2))$ again. $\quad\square$

NSN is interesting because it contains a 'pseudo-zero', the least infinite number m (and a pseudo-unity $m + 1$ as Richard Benham pointed out). That is, for any nonzero t, $\vdash t \times m = m$, and for any infinite t, $\vdash t + m = t$.

Now since m is a pseudo-zero, $m^3 + m^3 = m^3$ holds, that is $\vdash (m \times m \times m) + (m \times m \times m) = (m \times m \times m)$, a counterexample to Fermat's Last Theorem (FLT) in this structure. At the time of writing, FLT seems to be a good bet following Andrew Wiles' argument, though Wiles has indicated that one case remains open. The situation can be analysed as follows. Neither FLT nor $\sim FLT$ can be expressed in the present $\{+, \times\}$ language, since the capacity to express exponentiation fully is absent. However, each instance of both FLT and $\sim FLT$ can be considered in the language, since one can write, for example, $(xxxx) + (yyyy) = zzzz$. Thus, if any one of these held in a model for Standard Arithmetic, then FLT would be false; and if any one held in a model of Peano Arithmetic, then FLT would be unprovable in $P\#$. This motivates the introduction of the symbols 'FLT' and '$\sim FLT$' into the language, where the former is evaluated as the minimum of the values of its instances (that is semantically the quantifier $\forall$ is treated as generalised conjunction), and the latter evaluated as the maximum of its instances. More exactly, the value $\sim FLT$ is the maximum of the values of all equations of the form (x times itself n times) + (y times itself n times) = (z times itself n times), where x, y, z are nonzero numbers and n exceeds 2; while the value of FLT is the $RM3$-complement of the value of $\sim FLT$. Now by the construction of NSN, $(m \times m \times m) + (m \times m \times m) = m \times m \times m$ has the value B in NSN, so $\sim FLT$ is B or T in NSN, and thus holds in NSN. The following definition is now necessary.

Definition 3.3. For any subset S of $\mathcal{L}$, the *Routley* $*$ of S,
$S^* =df \{A :\sim A \text{ does not belong to } S\}$.

The Routley $*$ operation is important in the semantics of relevant logics, and is used in several places in this book. It is known that if Th is an inconsistent $RM3$-theory determined by an assignment I, then Th^* is an incomplete theory determined by taking I and changing only the set of designated values $\nabla = \{T\}$.

Proposition 3.4. FLT is true iff $\sim FLT$ is exactly B in NSN, and iff neither $\sim FLT$ nor FLT is in NSN^*; and FLT is false iff $\sim FLT$ is T in NSN, and iff $\sim FLT$ is in NSN^* and FLT is not in NSN^*.

Proof. That $\sim FLT$ is B in NSN iff neither $\sim FLT$ nor FLT is in NSN^*, follows from the well known fact that both A and $\sim A$ are in an $RM3$-theory Th iff neither A nor $\sim A$ are in Th^*. Now by an uncontroversial argument of Tarski, FLT is false iff for some finite x, y, z not classically identical with 0 and n not classical identical with 0, 1 or 2, $x^n + y^n = z^n$ holds in classical Robinson arithmetic. Hence, if $\sim FLT$ is exactly B in NSN, then there are no such finite x, y, z, n, and thus FLT is true. Conversely if FLT is true then there are no such finite x, y, z, n to raise the value of $\sim FLT$ to T in NSN, and $\sim FLT$ is exactly B. For the second part of the proposition, FLT is false iff these finite x, y, z, n exist, iff $\sim FLT$ is T in NSN. But also it is well known that A is T in Th iff A is in Th^* and $\sim A$ is not in Th^*. So $\sim FLT$ is T in NSN iff $\sim FLT$ is in NSN^* and FLT is not in NSN^*.□

Unfortunately, the job of proving $\sim FLT$ to be T in NSN seems to be no easier than finding a refuting instance to FLT, that is a finite x, y, z, n with $x^n + y^n = z^n$, *etc.* ever was.

CHAPTER 4: ORDER

1. Order and Equality without Function Symbols

So far, we have only been considering equational theories, that is, theories whose only primitive predicate symbol is $=$. We have been looking at how function symbols such as $+$ and $\times$ behave inconsistently in such theories. In this chapter a second primitive predicate $<$ is added to $=$, interpreted as 'less than'. In the first section, function symbols are omitted. The main result of this section is as follows. There is a well-known metatheorem of classical model theory to the effect that the theory of dense order without endpoints is $\aleph_0$-categorical, that is, that all classical models of that theory of cardinality $\aleph_0$ are isomorphic. We see that this breaks down for $RM3$-theories, given a natural extension of the notion of isomorphism of models to the more general case. It follows that the classical metatheorem depends on the assumption of classical logic. Such is not always the case: various other results in this book are invariant with respect to changes in background logic away from the classical case. In the next section, function symbols are re-introduced and several results about these are summarised.

The following axiomatises the classical theory of dense order without endpoints (see *e.g.* [22]).

(1) Irreflexivity $(x)(\sim x < x)$

(2) Asymmetry $(x,y)(x < y \supset \sim y < x)$

(3) Transitivity $(x,y,z)((x < y \& y < z) \supset x < z)$

(4) Comparability $(x,y)((\sim x = y \& \sim x < y) \supset y < x)$

(5) Exclusiveness $(x,y)((x = y \supset (\sim x < y \& \sim y < x) \& (x < y \supset \sim x = y))$

(6) No endpoints $(x)(\exists y,z)(x < y \& z < x)$

(7) Denseness $(x,y)(x < y \supset (\exists z)(x < z \& z < y))$

(8) Mixing $(x,y,z)(x = y \supset ((y < z \supset x < z) \& (z < y \supset z < x)))$

It is well known that all classical models of cardinality $\aleph_0$ of these axioms are isomorphic. To compare with $RM3$-models, a definition of isomorphism is needed which reduces to the usual in the classical case. The following seems adequate at least where every element of the domains is named.

Definition 4.1. Two models $\langle D_1, I_1 \rangle$ and $\langle D_2, I_2 \rangle$ are *isomorphic* iff there is a 1-1 correspondence $f : D_1 \rightarrow D_2$ such that for all names $t_1, \ldots t_n, t_{n+1}, \ldots t_{2n}$, if $I_2(t_{n+1}) = f(I_1(t_1))$ and $\ldots$ and $I_2(t_{2n}) = f(I_1(t_n))$ then for all atomic F, $F t_1 \ldots t_n$ holds in I_1 iff $F t_{n+1} \ldots t_{2n}$ holds in I_2.

Proposition 4.2. There are nonisomorphic $RM3$-models of the theory of dense order without endpoints.

Proof. For both models, take as names the rational numbers Q. For $\langle D_1, I_1 \rangle$ set $D_1 = Q$; set $I_1(t) = t$ for all t; set $I_1(t_1 = t_2) = T$ if $t_1 = t_2$ else $I_1 = F$; and set $I_1(t_1 < t_2) = T$ if $t_1 < t_2$ else $I_1 = F$. That is I_1 is the classical (and so $RM3$-) model of the $\{=, <\}$ theory of Q. For $\langle D_2, I_2 \rangle$ set $D_2 =$ the integers Z; set $I_2(t) =$ the integral part of t; set $I_2(t_1 = t_2) = B$ if $I_2(t_1) = I_2(t_2)$ else $I_2 = F$; and set $I_2(t_1 < t_2) = B$ if $I_2(t_1) \leq I_2(t_2)$ else $I_2 = F$. By the Extendability lemma, I_2 is an extension of I_1 and so (1)–(8) above hold in I_2. But there is no 1-1 correspondence between Z and Q which preserves atomic sentences of I_1 and I_2, since any correspondence eventually reverses the order on some elements. $\square$

This shows that the proof of $\aleph_0$-categoricity is not invariant with regard to background logic, but depends on the special properties of classical logic. Note also that I_2 is transparent; since if $t_1 = t_2$ holds then clearly for all atomic F, $I_2(F t_1) = I_2(F t_2)$, then use Proposition 2.13(1). The discreteness postulate $(x)(\exists y)(x < y \& (z)((x < z \& \sim y < z) \supset y = z))$ also holds in I_2 which shows that discreteness and denseness postulates can be jointly satisfied inconsistently.

2. Order and Equality with Function Symbols

There are many such models and theories with different properties. These are studied in [34]. Summarising those results:

(1) Primitive symbols $\{=, <, +, -, \times\}$; names for all the integers; domain $D =$ finite integers modulo m of Chapter 2; $I(t) = t \bmod m$; $I(+, -, \times) = (+, -, \times) \bmod m$; $I(t_1 = t_2) = B$ if $I(t_1) = I(t_2)$ else $I = F$; and $I(t_1 < t_2) = B$ if $t_1 \leq t_2$ else $I = F$. All sentences of the classical consistent complete theory of the ring of integers Z hold including the Sum Law $(x, y, z)(x < y \supset x + z < y + z)$ and Product Law $(x, y, z)(x < y \supset (0 < z \supset x \times z < y \times z))$. Functionality and hence transparency fail: $t_1 = t_2$ holds iff $t_1 \bmod m = t_2 \bmod m$; but $t_1 \bmod m = t_2 \bmod m$ plus $t_1 < t_3$ do not ensure $t_2 < t_3$ since t_2 might be too large. The equational subtheory is transparent.

(2) As for (1) but with $I(t_1 < t_2) = B$ if $I(t_1) \leq I(t_2)$ else $I = F$. These are transparent but not all sentences of the classical theory of Z with names hold, e.g. in mod 3, $I(2 < 4) = F$ because not 2 mod $3 \leq 4$ mod $3 = 1$.

(3) As for (1) but with $I(t_1 < t_2) = B$ all terms t_1, t_2. These are transparent and all sentences of the $\{=, <, +, -, \times\}$ theory of Z with names hold, but at the cost of triviality in the $\{<, +, -, \times\}$ subtheory.

(4) Primitive symbols $\{=, <, +, \times, /\}$. Take the two-element model with names for the rationals at the end of Chapter 2 Section 3 and add $I(t_1 < t_2) = B$ if $I(t_1) \leq I(t_2)$ else $I = F$. This is transparent and extends the classical consistent complete arithmetic and order theory of the nonnegative rationals.

(5) Primitive symbols $\{=, <, +, -, \times, /\}$; names for all the real numbers; $D = \{0, 1, \ldots, p - 1\}$; for any term t set $I(t) = 0$ if $t \leq 0$ or $t > p - 1$ else $I(t) =$ the greatest integer $\leq t$; set $I(t_1 = t_2) = B$ if $I(t_1) = I(t_2)$ else $I = F$; and set $I(t_1 < t_2) = B$ if $I(t_1) \leq I(t_2)$ else $I = F$. The theory is transparent. The $\{=, <\}$ subtheory extends that of the classical consistent complete real numbers R, including the continuity schema ([50], p.31). The $\{=, +, -, \times, /\}$ subtheory extends the classical theory of fields. Also Sum and Product laws hold. It is not known whether all classical consequences of field axioms + (1)–(8) + continuity scheme hold.

(6) $\{=, <, +, -, \times, /\}$; names for all real numbers; $D = \mathrm{mod}\,p$; $I(+, -, \times, /\}$ are mod p; set $I(t) = 0$ if $t \leq 0$, $I(t) =$ the least whole number $\geq t$ if $0 < t \leq p-2$, else $I(t) = p-1$; set $I(t_1 = t_2) = B$ if $I(t_1) = I(t_2)$ else $I = F$; and set $I(t_1 < t_2) = T$ if $I(t_1) < I(t_2)$, $I(t_1 < t_2) = B$ if $I(t_1) = I(t_2)$ else $I = F$. These are transparent, and the $\{=, <\}$ subtheory extends that of classical R, and the $\{=, +, -, \times, /\}$ subtheory extends that of classical fields. But Sum and Product laws fail, and in different moduli. (Sum law: in mod 2 or mod prime $p \geq 3$, $p-2 < p-1$ holds but $(p-2) + 1 < (p-1) + 1$ does not hold. Product law: in mod $p \geq 3$, $1 < p-1$ and $0 < p-1$ are both T but $I(1 \times (p-1) < (p-1) \times (p-1)) = I(p-1 < 1) = F$.

Finally, inconsistency can be isolated to subtheories containing one primitive predicate and not the other, while retaining transparency. In mod p, both the following are transparent:

(7) $I(t_1 = t_2) = B$ if $I(t_1) = I(t_2)$ else $I = F$, and $I(t_1 < t_2) = T$ if $I(t_1) < I(t_2)$ else $I = F$.

(8) $I(t_1 = t_2) = T$ if $I(t_1) = I(t_2)$ else $I = F$, and $I(t_1 < t_2) = B$ if $I(t_1) \leq I(t_2)$ else $I = F$.

CHAPTER 5: CALCULUS

1. Introduction

As noted in Chapter 1, there have been calls recently for inconsistent calculus, appealing to the history of calculus in which inconsistent claims abound, especially about infinitesimals (Newton, Leibniz, Bernoulli, l'Hospital, even Cauchy). However, inconsistent calculus has resisted development. There seem to be at least two reasons for this. First, as we have seen, the functional structure of fields interacts with inconsistency to produce triviality even in the purely equational part of theories, in a way which normal paraconsistentist contradiction-containment devices, such as weakening *ex contradictione quodlibet*, do not prevent. Stronger theories, including set membership, terms of infinite length, order, limits and integration, are then infected with the same triviality. Second, the functional structure of inconsistent set theory remains difficult to control, and seems to require sacrifice of logical principles in addition to, and more natural than, ECQ. (See Meyer *et.al.* [29], Slaney [53], but also below Chapter 14.) But unless there are distinctive inconsistent theories of the order of strength of classical analysis, then the claim that the history of the calculus supports paraconsistency is undermined. Inconsistency might well instead be a symptom of confusion.

This chapter extends inconsistency to the case of inconsistent equational theories strong enough for a reasonable notion of differentiation of polynomials, in order to show that inconsistency does not cripple such an equational differential calculus. It turns out to be instructive to begin not with an inconsistent theory but with an incomplete consistent (intuitionist) theory; which can be seen to have some similarities with, and advantages over, the well-known intuitionist theory Synthetic Differential Geometry (SDG). It also has the advantage of showing how incomplete theories are just as amenable to treatment by these methods as inconsistent theories are. In section 2 a congruence relation is defined on the noninfinite hyperreal numbers, and the algebra of equivalence classes so obtained is shown to have the structure of a nilpotent ring. This allows the functional properties of the incomplete theory to be defined in section 3. In section 4 the calculus of polynomials is described and results on incompleteness, nilpotence, Taylor formulae, polynomial differentiation and continuity are obtained. These are compared with SDG, and the similarities, advantages (mostly simplicity) and limitations of the comparison are discussed. In section 5 it is shown that a small change allows a very similar inconsistent theory to be defined. Section 6 deals with integration, and in the final section, various further directions are sketched. On the basis of these results it is argued that the fact that the same functional structure can underlie inconsistent, incomplete, or classical theories suggests that the functional aspects of mathematics are more important than squabbles at the sentential level over ECQ, inconsistency, incompleteness, *etc.*

2. A Nilpotent Ring of Hyperreal Numbers

We begin with the usual classical arithmetic of the field of hyperreal numbers R^*, with operations $\{+, -, \times, /\}$. Any hyperreal number has a representation $H+r+d$ where H is an infinite number, r a real number and d an infinitesimal, the reciprocal of an infinite number. If two hyperreal numbers x, y are at most infinitesimally distinct, we write $x \approx y$. Thus any infinitesimal ≈ 0. The subfield of real numbers is called R. For each nonzero x in R^*, the binary relation $\approx x$ is defined by $x_1 \approx x$ $x_2 = df$ (x_1/x) is at most infinitesimally different from (x_2/x); that is $(x_1/x) \approx (x_2/x)$, that is $(x_1 - x_2)/x$ is infinitesimal, written $(x_1 - x_2)/x \approx 0$. For fixed x this is an equivalence relation on R^*, as is easy to verify. It is not however a congruence. For example if $(x_1 - x_2)/x$ is infinite w.r.t. x_3, then $x_1 \approx x$ x_2 does not ensure $(x_1/x_3) \approx x$ (x_2/x_3). However, if x is any infinitesimal δ, then a congruence on the *noninfinite* hyperreal numbers, w.r.t. the operations $\{+, -, \times\}$, is obtained; as well as an associated ring of equivalence classes. So, fix δ; then we can define:

Definition 5.1. Let S be the set of noninfinite hyperreals, that is of the form $x + d$ where x is any real number and d any infinitesimal (possibly 0). Then

$D =df \{d \in S : \text{for some positive integer } k, d^k/\delta \approx 0\}$;

$S- =df\ S$ with d restricted to D.

Proposition 5.2. The relation $\approx\delta$ is a congruence on S and on $S-$.

Proof. Let $(x_1 + d_1) \approx\delta\ (x_2 + d_2)$, that is $((x_1 + d_1) - (x_2 + d_2))/\delta \approx 0$; and let $(x_3 + d_3) \approx \delta\ (x_4 + d_4)$, that is $((x_3 + d_3) - (x_4 + d_4))/\delta \approx 0$. Then $(((x_1 + d_1) + (x_3 + d_3)) - ((x_2 + d_2) + (x_4 + d_4)))/\delta \approx 0$, that is $((x_1 + d_1) + (x_3 + d_3)) \approx\delta\ ((x_2 + d_2) + (x_4 + d_4))$.

The subtraction case is the same. For multiplication note that $(x_1+d_1) \approx\delta\ (x_2+d_2)$ iff $x_1 = x_2$ and $d_1/\delta \approx d_2/\delta$. Now

$$((x_1 + d_1) \times (x_3 + d_3)) - ((x_2 + d_2) \times (x_4 + d_4)))/\delta$$
$$= (x_1x_3 - x_2x_4 + x_3d_1 - x_4d_2 + x_1d_3 - x_2d_4 + d_1d_3 - d_2d_4)/\delta.$$

The first pair of terms cancel. The second pair are $\approx$ since multiplication by a real number does not disturb $\approx$, so their difference is infinitesimal. Ditto the other two pairs in the sum, so the whole sum is infinitesimal. This shows that $\approx\delta$ is a congruence for $\{+, -, \times\}$ on S. For $S-$, suppose that

$$d_1^{k_1}/\delta \approx d_2^{k_2}/\delta \approx d_3^{k_3}/\delta \approx d_4^{k_4}/\delta \approx 0.$$

The infinitesimal part of $(x_1 + d_1) + (x_3 + d_3)$ is $(d_1 + d_3)$, and

$$(d_1 + d_3)^{k_1+k_3}/\delta = \left(\sum_{i=0}^{k_1+k_3} (k_{1_i} + k_3)d_1^{k_1+k_3-i}d_3^i \right) /\delta.$$

But each term of the sum ≈ 0 so the whole sum is; and so $(x_1 + d_1) + (x_3 + d_3)$ is in $S-$, as is obviously $(x_2 + d_2) + (x_4 + d_4)$. Subtraction is similar. For multiplication $(x_1 + d_1) \times (x_3 + d_3) = (x_1x_3 + x_1d_3 + x_3d_1 + d_1d_3)$. Now

$$0 \approx x_1d_3^{k_3} \approx x_3d_1^{k_1} \approx d_1d_3^{\min(k_1,k_3)},$$

so each term of the sum is in $S-$; so the whole sum is as in the proof of the addition case above.   □

Note that the proof of congruence breaks down for the case of division, for example if all x_i are 0 and $d_1 - d_2 = d_3$. It follows from Proposition 5.2 that the set of equivalence classes under $\approx\delta$ form a ring (call it $\mathbb{R}$) w.r.t. the induced operations $\{+, -, \times\}$. Denote the equivalence class of any element $x + d$ by $[x + d]$. $\mathbb{R}$ has the following properties.

Proposition 5.3. (1) For any real numbers x_1, x_2, $[x_1] = [x_2]$ iff $x_1 = x_2$.
(2) For any infinitesimals d_1, d_2, if $\left[d_1^2\right] = \left[d_2^2\right] = [0]$ then $[d_1] \times [d_2] = [0]$.
(3) For any nonnegative integer k, there is some infinitesimal d with $[d^{k+1}] = [0]$ but not $[d^k] = [0]$.

Proof. (1) If x_1, x_2 are real, then not $(x_1 - x_2)/\delta \approx 0$ unless $x_1 = x_2$.
(2) Let $d_1^* = d_1^2/\delta$ and $d_2^* = d_2^2/\delta$. By hypothesis, $d_1^* \approx 0 \approx d_2^*$. But

$$d_1 d_2/\delta = \left(d_1^2 d_2^2/\delta^2\right)^{1/2} = \left(d_1^*\right)^{1/2}\left(d_2^*\right)^{1/2},$$

which is infinitesimal if d_1^* and d_2^* are.
(3) Consider $\delta^2, \delta, \delta^{1/2}, \delta^{1/3}, \ldots, etc.$.   □

Proposition 5.4. For any infinitesimal δ and any positive integer k, there is an infinitesimal d such that d^{k+1}/δ is infinitesimal while d^k/δ is infinite.

Proof. Let $d = \delta^{(k+1)/k(k+2)}$. Now

$$d^{k+1}/\delta = \delta^{(k+1)^2/k(k+2)}/\delta = \delta^{(k^2+2k+1)/k(k+2)}/\delta^{(k^2+2k)/(k^2+2k)} = \delta^{1/(k^2+2k)} \approx 0.$$

But,

$$d^k/\delta = \delta^{k(k+1)/k(k+2)}/\delta = \delta^{(k+1)/(k+2)}/\delta^{(k+2)/(k+2)} = \delta^{-1/(k+2)} = 1/\delta^{1/(k+2)}$$

which is infinite.   □

Definition 5.5. $D_0 =_{df} [0]$; and, for all positive integers k,
$$D_k =_{df} \{[d] : [d^{k+1}] = [0] \text{ and not } [d^k] = [0]\}.$$
Note that $\{[d] : d \text{ is in } D\} = \cup_{\text{all } k}(D_k)$. (For D see Definition 5.1.)

Proposition 5.6. For all positive integers k, (1) there is a $[d]$ in D_k such that for all $[d_1]$ in D, $[d_1] \times [d^k] = [0]$; and (2) there is a $[d]$ in D_k and a $[d_1]$ in D_{k+2}

such that not $[d_1] \times [d^k] = [0]$.

Proof. (1) Let d be $\delta^{1/k}$. Now $d^k/\delta = 1$ not ≈ 0. But $d^{k+1}/\delta = 1.\delta^{1/k} \approx 0$. Hence $[d]$ is in D_k. But also, for any infinitesimal d_1, $d_1.d^k/\delta = d_1 \approx 0$; so that $[d_1] \times [d^k] = [0]$.

(2) Let d be $\delta^{(k+1)/k(k+2)}$ as in Proposition 5.4, and let d_1 be $\delta/(d^k)$. Now by the argument of Proposition 5.4, $[d]$ is in D_k. Further,

$$d_1 = \delta/d^k = \delta/\delta^{(k+1)/(k+2)} = \delta^{1/(k+2)} .$$

So

$$d_1^{(k+2)}/\delta = \delta^{(k+2)/(k+2)}/\delta^{(k+2)/(k+2)} = 1 \text{ not } \approx 0;$$
$$\text{and} \quad d_1^{(k+3)}/\delta = \delta^{(k+3)/(k+2)}/\delta^{(k+2)/(k+2)} = \delta^{1/(k+2)} \approx 0.$$

Hence d_1 is in D_{k+2}. Finally, $(d_1 d^k)/\delta = 1$ not ≈ 0; so that not $[d_1] \times [d^k] = [0]$. $\square$

Definition 5.7. An element d of an algebra is *nilpotent of degree k* if $d^{k+1} = 0$; and *strictly nilpotent of degree k* if $d^{k+1} = 0$ but not $d^k = 0$; and an algebra is *(strictly) nilpotent of degree k* if it has (strictly) nilpotent elements of degree k.

Proposition 5.3(1) shows that $\mathbb{R}$ has a subfield isomorphic to the real numbers R. This field of equivalence classes will also be referred to as R. Now in $\mathbb{R}$ we can as usual write $[x]^k$ for $[x^k]$ and drop the multiplication signs or use dots. Proposition 5.3(3) shows that $\mathbb{R}$ is strictly nilpotent of all degrees. Proposition 5.3(2) is relevant to the comparison with SDG in section 4. While all elements of D_k go to zero on being raised to the $k + 1$st power and not for any lesser power, Proposition 5.6 shows that these elements fall into two classes: those whose kth power multiplied by any nilpotent element goes to zero, and those whose kth power has a nonzero product with some nilpotent element. This is also relevant to section 4.

3. An Incomplete Theory

This section specifies an incomplete model based on the three-valued intuitionist logic $J3$. In the following section theorems of calculus are proved in it.

Definition 5.8. The theory CJ is specified by:

(1) background logic $J3$ (Three values $\{F, N, T\}$ with $\nabla = \{T\}$. See Definition 2.9.);

(2) names for all noninfinite hyperreal numbers;

(3) term forming operators $\{+, -, \times\}$;

(4) the single binary relation $=$;

(5) sentential operators $\{\sim, \&, \vee, \rightarrow\}$;

(6) Two sorts of object language variables, each with several sorts of associated quantifiers

    (6i) variables $x, x_0, x_1, \ldots$ and two associated pairs of quantifiers $(\forall \in \mathbb{R})$, $(\exists \in \mathbb{R})$ and $(\forall \in R)$, $(\exists \in R)$, and

    (6ii) variables $d, d_0, d_1, \ldots$ and associated pairs of quantifiers $(\forall \in D)$, $(\exists \in D)$ and for every positive integer k, $(\forall \in D_k)$, $(\exists \in D_k)$;

(7) $\{\supset, \equiv, \leftrightarrow\}$ are defined in the usual way, $(E!x \in R)(Fx)$ is defined as $(\exists x \in R)(Fx \& (\forall x_0 \in R)(Fx_0 \rightarrow x = x_0))$;

(8) The model $\langle D, I \rangle$ is specified by

    (8i) $D = \mathbb{R}$

    (8ii) For every name t, $I(t) = [t]$

    (8iii) $I(+, -, \times)$ are the corresponding ring operations on $\mathbb{R}$

    (8iv) For any terms t_1, t_2, set $I(t_1 = t_2) = T$ if $I(t_1) = I(t_2)$, set $I(t_1 = t_2) = N$ if $I(t_1) \neq I(t_2)$ but the hyperreal number $(t_1 - t_2)/\delta$ is noninfinite, else set $I(t_1 = t_2) = F$.

(9) For every quantified sentence of the form $(\forall v \in X)Fv$, $I((\forall v \in X)Fv) = glb\{y : \text{for some term } t, I(t) \text{ is in } X \text{ and } I(Ft) = y\}$; and $I((\exists v)Fv) = lub\{\text{the same set}\}$, where v is any variable and X is $\mathbb{R}, R, D$ or D_k;

(10) CJ is then $\{A : I(A) = T\}$.

The model just described is transparent. (This follows from the facts (1) that $t_1 = t_2$ holds iff $I(t_1) = I(t_2)$, and hence (2) that if $t_1 = t_2$ holds then $I(Ft_1) = I(Ft_2)$ for any atomic F. The latter then serves as the base of an obvious induction for all F.) This means that there is full functionality for calculation, so that advantage can be taken of facts about nilpotence such as $\vdash \delta^2 = 0$ for simplifying calculation. Elimination of 'second order' terms in a series has looked attractive from as long ago as Newton. The theory CJ is intuitionist in the senses (i) that $J3$ is three-valued intuitionist logic, and (ii) that CJ is incomplete: since if $I(\delta = 0) = N$ then $I(\sim \delta = 0) = F$, so that neither $\vdash \delta = 0$ nor $\vdash \sim \delta = 0$, although $\vdash \sim\sim \delta = 0$ & $\sim \delta^{1/2} = 0$. Finally note that the wholly classical two-valued theory of $\mathbb{R}$ can be obtained by taking $I(t_1 = t_2) = T$ if $I(t_1) = I(t_2)$ else $I = F$. This shows that classical two-valued model theory can be obtained as a special case.

4. Incomplete Differential Calculus

In this section it is shown that Taylor's formula and polynomial differentiation laws hold in CJ. A definition of limits can be given, and it is proved that every function is continuous. It is shown that the theory has some similarities with a corresponding part of Synthetic Differential Geometry, and the dissimilarities are outlined.

Definition 5.9. A *functional expression* (abbreviated to *function*) is the result of replacing any term or terms inside any term, by variables. A function with no remaining names denoting infinitesimals is called a *real* function. If f is a function with a single free variable v (possibly occurring more than once) then this is indicated by $f(v)$. If v_1 and v_2 are variables of any sort, then $f(v_1 + v_2)$ is the result of replacing v by $v_1 + v_2$ throughout; and if t_1, t_2 are any terms then $f(t_1 + t_2)$ is the result of replacing v by $t_1 + t_2$ throughout. Similarly for $-$ and $\times$. $(E!x_1, \ldots, x_k \in R)$ is defined as $(E!x_1 \in R)\ldots(E!x_k \in R)$. (See Definition 5.8 (7).)

Proposition 5.10. If $f(x)$ is any real function, then for all positive integers k,

$$\vdash (\forall x \in R)\,(E!x_1, \ldots, x_k \in R)\,(\forall d \in D_k)\left(f(x + d) = f(x) + x_1 d + \ldots + x_k d^k\right).$$

Proof. If $f(x)$ is a real function, then by the polynomial laws of R^*, for any term t, $I(f(t))$ is identical with $I(t_0 + t_1 t + \ldots + t_n t^n)$, where the t_i are names denoting real numbers. This is clear because identities are not destroyed in passing from R^*

to $\mathbb{R}$. So we may restrict attention to functions of the form $t_0 + t_1 x + \ldots + t_n x^n$ where the t_i are names for real numbers so that the $I(t_i)$ are in R. We abbreviate these functions by $\sum_{i=0}^{n} t_i x^i$. Then for any such $f(x)$ and any term t from R and any term d with $I(d)$ in some D_k, $f(t + d)$ is $t_0 + t_1(t + d) + \ldots + t_n(t + d)^n$. So $I(f(t+d)) = I(t_0) + (I(t_1))(I(t) + I(d)) + \ldots$ etc. The operations on the right hand side of the last expression obey the polynomial laws, so that sum can be computed using the binomial expansion. If $n \leq k$, the nilpotence of the element d does not affect this expansion, and (α) below follows by normal arithmetic. If $n > k$, those terms of the binomial expansion of $I(f(t + d))$ which contain $[d^{k+1}]$ as a factor are identical with $[0]$. So $I(f(t + d))$ computes to

$$I\left(t_0 + t_1 t + \ldots + t_n t^n\right) + I\left(\left(\sum_{i=1}^{n} \binom{i}{1} t_i t^{i-1}\right) d\right) + \ldots + I\left(\left(\sum_{i=1}^{n} \binom{i}{k} t_i t^{i-k}\right) d^k\right) \quad (\alpha)$$

Hence by the assignment rules for quantifiers

$$\vdash (\forall x \in R)(\exists x_1 \ldots x_k \in R)(\forall d \in D_k)\left(f(x + d) = f(x) + \sum_{i=1}^{k} x_i d^i\right) \quad (\beta)$$

The next part of the argument, for uniqueness, uses the postulate that the $I(t_i)$ are real. We need to conjoin to (β) the following:

$$(\forall x_{k+1} \ldots x_{2k} \in R)((\forall d \in D_k)\left(f(t + d) = f(t) + \sum_{i=1}^{k} x_{k+i} d^i\right)$$
$$\rightarrow \left((t_1 = x_{k+1})\& \ldots \&(t_k = x_{2k})\right),$$

where the t_i are a relabelling of the coefficients of (α). Eliminating quantifiers to appropriately assigned terms, we need to prove that:

$$\vdash (\forall d \in D_k)\left(f(t + d) = f(t) + \sum_{i=1}^{k} t_{k+i} d^i\right) \rightarrow \&_{i=1}^{k} (t_i = t_{k+i}) \quad (\gamma)$$

If the consequent takes the value T then (γ) holds by the tables for $\rightarrow$. If the consequent does not take the value T then there are two cases: either (i) $t_k = t_{2k}$ does not hold, or (ii) some other $t_i = t_{k+i}$ does not hold.

(case i:) If $t_k = t_{2k}$ does not hold, then $I(t_k) \neq I(t_{2k})$. Now since t_k and t_{2k} are real, $(t_k - t_{2k})/\delta$ is infinite, so $I(t_k - t_{2k}) = F$. But by Proposition 5.4, there is some infinitesimal hyperreal number d such that d^k/δ is infinite; hence $(t_k - t_{2k})d^k/\delta$ is infinite. If every other $t_i = t_{k+i}$ holds, then $[t_i] = [t_{k+i}]$ and $t_i = t_{k+i}$ in R. So in R^*,

$$f(t+d) - \left(f(t) + \sum_{i=1}^{k} t_{k+i} d^i\right) = f(t) + \sum_{i=1}^{k} t_i d^i - \left(f(t) + \sum_{i=1}^{k} t_{k+i} d^i\right) = (t_k - t_{2k})d^k .$$

But the latter is infinite w.r.t. δ. So in $\mathbb{R}$,

$$I(f(t+d)) \neq I\left(f(t) + \sum_{i=1}^{k} t_{k+i} d^i\right).$$

But also in R^*,

$$\left(f(t+d) - \left(f(t) + \sum_{i=1}^{k} t_{k+i} d^i\right)\right) / \delta$$

is infinite. Hence the antecedent of (γ) is F and (γ) holds by the table for $\rightarrow$.

(case ii:) Otherwise, let i be the least integer for which $t_i = t_{k+i}$ does not hold. Then choosing the same d, in R^* we have

$$f(t+d) - \left(f(t) + \sum_{i=1}^{k} t_{k+i} d^i\right) = (t_i - t_{k+i})d^i + \text{ higher powers of } d.$$

But the first term is infinite w.r.t. δ if d^k is. So, as in case (i), in $\mathbb{R}$,

$$I(f(t+d)) \neq I\left(f(t) + \sum_{i=1}^{k} t_{k+i} d^i\right).$$

But in R^*,

$$\left(f(t+d) - \left(f(t) + \sum_{i=1}^{k} t_{k+i} d^i\right)\right) / \delta$$

is infinite. Hence again the antecedent of (γ) is F and so (γ) holds. $\qquad\square$

Consider the case $k = 1$. Then for any $[d]$ in D_1 and any real t, $\vdash f(t+d) = f(t) + t_1 d$, for some term t_1 with $I(t_1)$ in R.

Definition 5.11. A function $g(x)$ is called a *derivative of* $f(x)$, if for any d in D_1 and any t with $I(t)$ in R, $\vdash f(t+d) = f(t) + d.g(t)$. If $g(x)$ is a derivative of $f(x)$, it is also denoted by $f'(x)$.

It is clear independently from classical real number calculus that there is always at least one derivative for each real function $f(x)$. So we have:

Proposition 5.12. (Taylor Formula) For any derivative $f'(x)$,

$\vdash f(t + d) = f(t) + d.f'(t)$; or $\vdash (\forall x \in R)(\forall d \in D_1)(f(x + d) = f(x) + d.f'(x))$.

Definition 5.13. An *n-th degree polynomial in the indeterminate* x is any function of the form $t_0 + t_1 x + \ldots + t_n x^n$, where the t_i are names, that is $\sum_{i=0}^{n} t_i x^i$.

Proposition 5.14. (Polynomial Differentiation) If f is any polynomial of the form $\sum_{i=0}^{n} t_i x^i$ with real coefficients t_i and $f'(x)$ is any derivative of f, then

$\vdash (\forall x \in R)(f'(x) = \sum_{i=1}^{n} i t_i x^{i-1})$.

Proof. From the Taylor formula, $\vdash (\forall x \in R)(f(x + d) = f(x) + d.f'(x))$ where $I(d)$ is in D_1. Hence $I(f(t + d)) = I(f(t)) + I(d).I(f'(t))$ for any term t with $I(t)$ in R. But $I(f(t + d)) = I(\sum_{i=0}^{n} t_i(t + d)^i)$. As in Proposition 5.10, this computes to $I(\sum_{i=0}^{n} t_i t^i) + \left(I\left(\sum_{i=1}^{n} \binom{i}{1} t_i t^{i-1}\right).I(d)\right) + \left(I\left(\sum_{i=2}^{n} \binom{i}{2} t_i t^{i-2}\right).I(d^2)\right) +$ higher powers of d. Since $I(d^2) = I(d^3) = \ldots = [0]$, all of these can be dropped. Thus we have

$$I(f(t + d)) = I(f(t)) + (I(d).I(f'(t)))$$

and also

$$= I(f(t)) + \left(I(d).I\left(\sum_{i=1}^{n} \binom{i}{1} t_i t^{i-1}\right)\right).$$

So since subtraction is one of the congruence operations,

$$I(d).I(f'(t)) = I(d).I\left(\sum_{i=1}^{n} \binom{i}{1} t_i t^{i-1}\right).$$

But since $I(d)$ is in D_1 and $I(t)$ and $I(t_i)$ are in R, this can only happen if

$$I(f'(t)) = I\left(\sum_{i=1}^{n} \binom{i}{1} t_i t^{i-1}\right).$$

But t was arbitrarily chosen from R. Hence $\vdash (\forall x \in R)(f'(x) = \sum_{i=1}^{n} i t_i x^{i-1})$. □

Definition 5.15.

(1) $\lim_{x \to t} f(x) = t_1 =df (\forall d \in D)(f(t + d) = t_1 \lor (\exists d_1 \in D)(f(t + d) - t_1 = d_1))$

(2) f is continuous at $t =df \lim_{x \to t} f(x) = f(t)$

(3) f is continuous $=df (\forall x \in R) (f$ is continuous at $x)$.

A definition of one sided limits can be given, but that is not done here because of the following proposition. (See also Chapter 6.) It is also noted that in the above definition of limit, the case where not $\vdash f(t) = t_1$ does not arise, as the following proposition shows.

Proposition 5.16. For every real function $f(x)$, $\vdash f$ is continuous.

Proof. It has to be proved, for every real term t, that:

$\vdash (\forall\, d \in D)(f(t + d) = f(t) \vee (\exists d_1 \in D)(f(t + d) - f(t) = d_1))$. But it follows from Proposition 5.10 that $\vdash (\forall\, d \in D_k)(f(t + d) = f(t) + t_1 d + \ldots + t_k d^k)$. If not all the real t_i are $= 0$, then $\vdash f(t + d) - f(t) = t_1 d + \ldots + t_k d^k$. It is obvious that raising the RHS to the power k is not (considered as a hyperreal number) infinitesimal w.r.t. δ (since its first term is not); while raising the RHS to the power $k + 1$ is infinitesimal w.r.t. δ (since each term is). Hence the RHS is in D_k. Thus $\vdash (\exists d_1 \in D)(f(t + d) - f(t) = d_1)$. The result follows by disjoining the alternatives and universal generalisation. □

Synthetic Differential Geometry (SDG), as expounded in Kock [18] (see also Bell [3]), is likewise an incomplete theory, with neither $\delta = 0$ nor $\sim \delta = 0$ holding. The theory of [18] has nilpotent elements of all degrees, while the theory of [3] concentrates on D_1. Neither proceeds from a construction on the classical hyperreal numbers, nor uses three valued model theory. In these theories, also, every function is continuous. The method of obtaining derivatives from the Taylor formula as in Proposition 5.12 is similar to that in [18], and is a variant of the usual classical treatment. Like SDG, Propositions 5.10 and 5.14 use the calculatory advantages of nilpotent elements, since these ensure that higher order differentials can ultimately be ignored.

The case $x = 0$ of Proposition 5.10 is Axiom $1'$ of [18], with the proviso that R in Proposition 5.10 is replaced by the whole domain there. If however R is replaced by $\mathbb{R}$ in Proposition 5.10 then it fails, as follows. Choose any d_1 in D_1 and let $f(x)$ be the function $d_1 x$. Then certainly $\vdash (\exists x \in R)(\forall\, d \in D_1)(f(d) = f(0 + xd))$, the

x in question being d_1. However, this x is not unique: for any other d_2 in D we have $\vdash (\forall\, d \in D_1)(d.d_2 = d.d_1 = 0)$ while not $\vdash d_1 = d_2$; so that the antecedent of $(\forall\, d \in D_1)(f(d) = f(0) + d.d_2) \rightarrow d_1 = d_2$ holds while the consequent does not hold. Indeed, f could even have a noninfinitesimal coefficient, $f(x) = (r + \delta)x$ say. For then the coefficient fails to be unique, since $\vdash (\forall\, x \in D_1)((5 + \delta)d = 0 = (5 + 2\delta)d)$ while not $\vdash 5 + \delta = 5 + 2\delta$. Thus the present theory is a theory of functions with real slopes as in classical nonstandard analysis.

The essential difference with SDG is that the D_1 part of the domain is postulated in SDG to contain elements d_1, d_2 such that not $\vdash d_1 d_2 = 0$, while in the present model this is not so (Proposition 5.3(2)). Correspondingly there fails the SDG Cancellation Principle $(\forall\, d \in D_1)(d.t_1 = d.t_2) \rightarrow t_1 = t_2$; for example when $I(t_1) = [\delta]$ and $I(t_2) = [2\delta]$ then the antecedent is T and the consequent is N. However, the Cancellation Principle holds for cases where the difference between $I(t_1)$ and $I(t_2)$ is infinite w.r.t. δ if they are different at all, such as the real numbers. For example, $\vdash (\forall\, x_1 x_2 \in R)((\forall\, d \in D_1)(d.x_1 = d.x_2) \rightarrow x_1 = x_2)$.

The failure of the Law of Excluded Middle (LEM) is of interest. The account of [18] links it to the holding of the Cancellation Principle and the continuity of every function. However in the present theory it is rather independent of the functional part of the construction, since the latter can also produce a classical two-valued model (end of section 3). The same point pertains to the inconsistent theory of the next section. This does not show that the 'correct' description is that of classical logic, however; to the contrary it suggests that functionality is mathematically prior to sentential logic.

SDG in [18] uses the mathematical machinery of Cartesian closed categories, which is considerably stronger than that of equational theories. On the other hand there is some simplicity in presenting the ideas of incompleteness, nilpotence, differentiability, limits, continuity *etc.* within the framework of nonclassical model theory. Also, the present approach permits investigation of similar theories with different nonclassical background logics (see section 5). Another point is that while

[18] maintains that SDG is an *essentially geometric* treatment of analysis, it is interesting how close one can get to SDG with resources merely from algebraic number theory and model theory.

5. Inconsistent Differential Calculus

Definition 5.17. The theory CR is obtained by changing the definition of CJ in the following ways (1) background logic is $RM3$ with elements $\{F, B, T\}$ and designated elements $\nabla = \{B, T\}$, (2) $I(t_1 = t_2) = T$ if $t_1 = t_2$ considered as hyperreal numbers, else $I(t_1 = t_2) = B$ if $I(t_1) = I(t_2)$, else $I(t_1 = t_2) = F$.

There are a number of other options here, for example background logic $P3$ (this inconsistent theory can be called CP; see next two chapters), or dropping the first clause of (2). The latter produces a transparent theory, whereas CR is functional but not transparent. (Proof of functionality: By inspection, $t_1 = t_2$ holds iff $I(t_1) = I(t_2)$. But [] is a congruence, so if $I(t_1) = I(t_2)$ then $I(t(t_1)) = I(t(t_2))$. Hence if $t(t_1) = t_3$ holds then $I(t(t_1)) = I(t_3)$, so that $I(t(t_2)) = I(t_3)$, and so $t(t_2) = t_3$ holds. Disproof of transparency: $\vdash \delta^2 = 0$, but while $\vdash \sim \delta^2 = 0$, neither $\vdash \sim 0 = 0$ nor $\vdash \sim \delta^2 = \delta^2$.) This means that on the one hand calculations using the advantages of $\vdash \delta^2 = 0$ can be carried out, while on the other hand one does not have to submit to $\vdash \sim t = t$ for any term t, an improvement on earlier chapters. The loss of transparency does not appear a serious disadvantage: while it changes the logical properties of the theory, particularly which theories it extends, it does not affect calculation. Now it can be shown that the main Propositions of the incomplete theory can be reproved for the inconsistent theory.

Proposition 5.18. If $f(x)$ is any real function, then for every positive integer k, $\vdash (\forall x \in R)\, (E!x_1 \ldots x_k \in R)\, (\forall d \in D_k)\, \big(f(x + d) = f(x) + x_1 d + \ldots + x_k d^k\big)$.

Proof. The proof that $I(f(t + d))$ computes to (α) as in Proposition 5.10, is identical. To prove uniqueness, we need to prove (γ). If the consequent of (γ) is T, then (γ) holds. And for real coefficients t_i, t_{k+i}, one never has $I(t_i = t_{k+i}) = B$.

Hence consider the case $I(t_i = t_{k+i}) = F$. Then $I(t_i) \neq I(t_{k+i})$. But also $(t_i - t_{k+i})/\delta$ considered as a hyperreal number is infinite, since the numerator is real and nonzero. Hence as in Proposition 5.10, for some d with $[d]$ in D_k, $d^i(t_i - t_{k+i})/\delta$ is noninfinitesimal. So $I(d^i t_i) \neq I(d^i t_{k+i})$, and the antecedent of (γ) is F as required.   □

Proposition 5.19. If f is any polynomial of the form $\sum_{i=0}^{n} t_i x^i$ with real coefficients t_i, then $\vdash (\forall\, x \in R)\,(f'(x) = \sum_{i=1}^{n} it_i x^{i-1})$.

Proof. Similar to Proposition 5.14.   □

Proposition 5.20. For every real function f, $\vdash f$ is continuous.

Proof. Similar to Proposition 5.16.   □

To repeat an earlier point, inconsistent calculus is not being recommended as superior or truer, though its nilpotent elements have some of the calculatory advantages of SDG. The aim is only to show that it exists, that inconsistency permits a reasonable amount of calculus without collapse, and hopefully that inconsistent theories can be of mathematical interest.

6. Integration

This can be done in a similar way to the classical nonstandard account in Keisler [17]. The theory following is not much more complex than Keisler's; but on the other hand nilpotent elements appear not to convey any particular advantage, unlike in the case of differentiation. If there are any advantages at all, they may be that nilpotent elements allow 'smearing out zero' and a theory of delta functions as in Chapter 7.

Definition 5.21. (Keisler) Let $f(x)$ be a real function on an interval $\mathcal{I}$. An *area function for* f is a two placed function $A : R \times R \to R$ satisfying (i) Addition Property $A(a, b) = A(a, c) + A(c, b)$ for all $a \leq c \leq b$ in $\mathcal{I}$, and (ii) Rectangle

Property $m(b - a) \le A(a,b) \le M(b-a)$ for all $a \le b$ in $\mathcal{I}$, where $m = \min f$ on $\mathcal{I}$ and $M = \max f$ on $\mathcal{I}$. The *finite Riemann Sum* $\sum_a^b f(x)\Delta x$ is defined to be:

$$\sum_a^b f(x)\Delta x =_{df} f(x_0)\Delta x + f(x_1)\Delta x + \ldots + f(x_{n-1})\Delta x + f(x_n)(b - x_n),$$

where Δx is any positive real number, n is the maximum integer such that $a + n\Delta x \le b$, and $x_0 = a$, $x_1 = x + \Delta x, \ldots, x_n = x + n\Delta x$. The *infinite Riemann sum* $\sum_a^b f(x)dx$, where dx is any nonzero infinitesimal, is the nonstandard natural extension of the finite Riemann sum. Intuitively, it is the same sum as the finite sum except that there are an infinite number of terms, that is n becomes the largest hyperinteger such that $a + ndx \le b$. The *definite integral of f from a to b* $\int_a^b f(x)dx$ is the standard part of the infinite Riemann sum.

That these are well defined follows from the next Proposition.

Proposition 5.22. (Keisler)

(1) The infinite Riemannian sum is always a finite hyperreal number.

(2) The definite integral from a to b is independent of the size of the (nonzero) infinitesimal dx.

(3) The definite integral of f from a to b is the unique area function for f.

(4) (i) $\displaystyle\int_a^b c.dx = c.(b - a)$

(ii) $\displaystyle\int_a^b c.f(x)dx = c.\int_a^b f(x)dx$

(iii) $\displaystyle\int_a^b (f + g)dx = \int_a^b fdx + \int_a^b gdx$

(iv) $f \le g$ implies $\displaystyle\int_a^b fdx \le \int_a^b gdx.$

(5) $\displaystyle\int_a^b f'(x)dx = f(b) - f(a).$ $\qquad\qquad\qquad\qquad\qquad\qquad\qquad\square$

Moving to CR, functional expressions can be generalised to any string in which x is the sole variable, with the proviso that the interpretation $I(\text{string})$ is a partial function on the domain and $I(\text{string}(a)) = I(\text{string})(I(a))$. Now it is clear that (1)–(5) above hold for all infinitesimals d or dx, so in particular they hold for all infinitesimals d such that not $\vdash_{CR} d = 0$. But passing to CR preserves functional equality for functions restricted to such infinitesimals, provided that reference to hyperintegers is restricted to the metalanguage (infinitesimals such as δ^2 behave like zero in virtue of $\vdash \delta^2 = 0$ plus functionality). Hence we have

Proposition 5.22. (1)–(5) of the previous proposition hold of CR.   □

Further aspects of integration, such as indefinite integrals, the Second Fundamental Theorem, etc. can be dealt with in a similar way. In the special case of polynomial integration antiderivatives are easy to find directly, and so (1)–(5) can be verified directly.

7. Conclusion

The inconsistent theory here can be regarded as yet another approach to the idea of an 'infinitesimal microscope' (see [17],[54],[55]). A microscope with 'resolving power' δ can be said to be a theory which inconsistently identifies with zero and one another all quantities which are infinitesimal w.r.t. δ. One is unable to distinguish between quantities below this 'order of infinitesimality' or 'order of relative identity'; they have all one another's properties in common.

Further directions in which these ideas might be developed include inconsistent superreals (see [55]), inconsistent polynomial rings in one or more indeterminates, and introducing set membership (see Chapter 10). Finally, perhaps the present theories satisfy some of the inconsistent intuitions of the classical analysts; but even if not, inconsistent theories should be investigated.

CHAPTER 6: INCONSISTENT CONTINUOUS FUNCTIONS

1. Introduction

The idea that motion or change is an inconsistent process has, as is well known, a long history. Recent nice work by Graham Priest [46] suggests that an inconsistent account of motion and change is at least possible. It is a further matter whether it is true; and despite Priest's arguments, there does not seem to be a compelling reason for rejecting the existing consistent account from classical physics, which is mathematically both simple and elegant. Priest argues that the classical account has it that motion is being in different places at different times; whereas what he wants is an intrinsic account of motion according to which an instantaneous state ought to be unambiguously change or nonchange, independently of its (distance) relations to other states. Against this, one is inclined to argue that the relations are nonetheless present; that an account in which the relations alone carry the change is therefore inevitably simpler; and that being in different places at different times is surely necessary for motion, and more importantly (at least given a positive definite metric) sufficient as well.

As part of his account, Priest appeals to the Leibniz Continuity Condition (LCC). This condition is that whatever holds throughout an interval holds at its limits, and there is evidence that Leibniz held it. Now there is a technical problem with the principle as thus stated. Since any strictly monotonic continuous function takes throughout any interval values less than its value at the right hand endpoint of the interval, then applying the LCC gives that the function takes a value both less than and equal to itself at the endpoint. But since any point is the endpoint of some such interval, the function is both less than and equal to itself at all points (and greater than itself at all points as well, since whatever is greater than itself is less than itself as well, by symmetry).

Now, while Priest appeals to the LCC, not much of his account depends on it,

I think. On the other hand, the LCC should not be dismissed too quickly. The above problem depends on applying the LCC to the relation 'less than'. But not so much harm ensues if the LCC is applied only to equations, as we see in this chapter. In other words, it can be argued that the correct sphere of application of the LCC is at the fundamental level of being expressed by the basic physical Laws of Nature, and laws and boundary conditions of dynamical systems, since all of these are expressed in functional-equational form. Indeed, the application of the LCC to at least some discontinuous functions has the consequence that they can be treated as inconsistently continuous, in a manner outlined in this chapter; and that the natural logic arising from them is closed set logic.

The LCC serves to motivate the following account; however it is stressed that the theory stands independently of the LCC. We are concerned here with a special case, namely certain functions which from a classical point of view are not everywhere differentiable, but which can from an inconsistent point of view be regarded as having continuous derivatives. We see that the inconsistent derivative of such functions is continuous, given a natural extension of the meaning of the latter. Differentiating these in turn leads to delta functions, and in the next chapter an inconsistent account of these is proposed.

2. Functionality

Consider the continuous function $g(t) = k_1 t$ for all real numbers $t \leq 0$ and $g(t) = k_2 t$ for all $t > 0$, where k_1 and k_2 are classically different real numbers. The usual story about its derivative is that $g'(t) = k_1$ for all $t < 0$, and $g'(t) = k_2$ for all $t > 0$, but that $g'(0)$ does not exist; since the left hand limit of $g(\delta t)/\delta t$ as $\delta t \to 0^-$ is k_1, while the right hand limit as $\delta t \to 0^+$ is k_2, and $k_1 \neq k_2$. Inconsistently, however, there is no particular reason not to allow both $\sim k_1 = k_2$ and also $k_1 = k_2$. The latter $k_1 = k_2$ would ensure both that LH derivative=RH derivative so that g is differentiable at $t = 0$, and also that the derivative function is continuous at $t = 0$ and thus continuous everywhere.

Considering then the 'derivative function' $f(t) = k_1$ for all $t \leq 0$ and $f(t) = k_2$ for all $t \geq 0$, we have that $f(0) = k_1$ and $f(0) = k_2$; and, since f is a function, also $k_1 = k_2$ at $t = 0$. For this to amount to an inconsistency, it must also be that $\sim k_1 = k_2$ at $t = 0$. So a first consideration should be that enough of the arithmetic of the real numbers holds in the space of values of f at $t = 0$, that $\sim k_1 = k_2$ holds. One might insist that a necessary condition on any account which represents the *oddity* of what is happening at $t = 0$, be that $\sim k_1 = k_2$ represents the *norm*, with $k_1 = k_2$ an *extra abnormality* at $t = 0$. (Even the orthodox account recognizes abnormality after a fashion, in that declaring that g' does not exist at $t = 0$ is a kind of incompleteness.)

It is known, however, that if $k_1 = k_2$ where classically these are distinct real numbers, then there follows by purely $\{+, -, \times, /\}$ substitutions in classical real number identities, the undesirable conclusion that any real number a is identical with every other real number b. (Proof: $k_1 - k_1 = k_1 - k_1$. Substituting k_2 for k_1, $k_1 - k_1 = k_2 - k_1$. Classically, LHS $= 0$; so $0 = k_2 - k_1$. Classically $((k_2 - k_1) \times (b - a))/(k_2 - k_1) = ((k_2 - k_1) \times (b - a))/(k_2 - k_1)$. Substituting in LHS, $(0 \times (b - a))/(k_2 - k_1) = ((k_2 - k_1) \times (b - a))/(k_2 - k_1)$. Classically LHS $= 0$ and RHS $= b - a$, so $0 = b - a$. Classically $a + (b - a) = a + (b - a)$, so substituting 0 for $b - a$ in LHS, $a + 0 = a + (b - a)$. Classically LHS $= a$ and RHS $= b$, so $a = b$.)

Hence whatever is happening in the space of values of f at $t = 0$ has a reduced functional structure compared with that of the full field structure at t other than zero, or else the value space would have no structure at all at $t = 0$. The latter in turn would make it impossible to distinguish at $t = 0$ between an inconsistent jump from k_1 to k_2 and an inconsistent jump to any other k_3. One of the advantages of the present account is that this distinction can be sustained to a fair degree.

In the above argument, once one gets to $0 = k_2 - k_1$, if one then allows 'scale changes' by multiplying by a real number c, then one gets $0 = (k_2 - k_1) \times c$ which spreads rapidly to $a = b$. So it is reasonable to say that the value space at $t = 0$ lacks a multiplicative structure. But as we see there is no particular reason not to allow the extra additive consequences of $0 = k_2 - k_1$.

A natural account exploits the fact that for any real number r there is a unique integer a and a unique real number b such that $r = a.|k_2 - k_1| + b$ and $0 \leq b < |k_2 - k_1|$. So it is proposed to take the map $h : R \to [0, |k_2 - k_1|)$ with $h(a.|k_2 - k_1| + b) = b$; and then at $t = 0$ to identify real-valued quantities c and d if $h(c) = h(d)$. This produces $0 = k_2 - k_1$ and $k_1 = k_2$ among other things. One can think of the space of values of f as undergoing an 'instantaneous slip' from k_1 to k_2 at $t = 0$. Another analogy is for an 'instantaneous cylindrification' of the value space, in which the whole positive and negative axes are wound respectively clockwise and anticlockwise around the finite halfopen interval $[0, |k_2 - k_1|)$.

Now the halfopen interval $[0, |k_2 - k_1|)$ has a natural additive structure, defined by $a +' b =_{df} h(a+b)$, the latter $+$ being real number sum. It is known that $+'$ so defined is a function on $[0, |k_2 - k_1|)$. This makes h an additive group homomorphism, and it is seen later that in the inconsistent theory holding at $t = 0$, functionality of $\{+, -\}$ is preserved. Coming at it from another direction, let $a \cong b =_{df} a - b$ is an integral multiple of $|k_2 - k_1|$, where a and b are real numbers. It is not difficult to prove that $\cong$ is a congruence w.r.t. addition ([4], p.148). Hence the map $h : r \to$ (the unique b such that $r \cong b$ and $0 \leq b < |k_2 - k_1|$) takes the additive group of real numbers to the additive group on $[0, |k_2 - k_1|)$ with $a +' b =_{df} h(a + b)$ and $-'a =_{df} |k_2 - k_1| - a$.

It is obviously important that there be at least some functions such as $\{+, -\}$ in the value space of the derivative function; otherwise it has no structure save identities and their denials. Quite a lot is definable in the additive group, for example all integer multiplications. However the corresponding definition of unrestricted multiplication fails of functionality, and so multiplication is left 'undefined' at $t = 0$. (In the next section this is dealt with logically using incompleteness.) This looks bad only if one forgets the origin of differentiation in the gradients of a scalar field, or quantity spaces or phase spaces. There, one might insist, congruences of difference, or metrical distance on a quantity scale, are absolute. This defines 'twice, thrice,

...the distance' on the difference as absolute. On the other hand, expansion or contraction of the value space by multiplying quantities by an un-unitted real number is merely a *scale* change and not a quantity change. That arithmetical laws might be different from time-to-time is easier intuitively to attribute to quantity spaces than to (apparently) universal real numbers. But it is not really an extra problem to deal formally with the value space of f as real numbers, rather than unitted quantities as here. One simply treats the situation formally as here, forgetting the implicit units following the values of g, g' and f.

3. Logic

To make this logically more precise, a language for describing the value space at various t and a background logic are needed. In the following it is convenient to think of the variable t of differentiation as ranging over times in accord with dynamical systems. It is seen later that the topology of the dynamical system described by the inconsistent function $f(t)$ provides its own logic.

Take as a language names for all real numbers, term forming operators $\{+, -, \times, /\}$, and a single binary predicate $=$. As background logic take the logic $PJ4$ (Definition 2.9), with four values $\{F, N, B, T\}$ and designated values $\nabla = \{B, T\}$. An *additive* term is one containing no occurrences of $\{\times, /\}$. Atomic sentences are assigned values by a function $I : \mathcal{L} \times Time \rightarrow \{F, N, T, B\}$ in accordance with

(1) For any time t other than 0 and any terms t_1, t_2, $I(t_1 = t_2, t) = T$ if $t_1 = t_2$ is true in classical real number theory, else $I(t_1 = t_2, t) = F$.

(2) If t_1 and t_2 are both additive terms then

    (2.1) $I(t_1 = t_2, 0) = T$ if $t_1 = t_2$ is true in classical real number theory; and

    (2.2) $I(t_1 = t_2, 0) = B$ if t_1 and t_2 are classically distinct real numbers but $h(t_1) = h(t_2)$, while $I(t_1 = t_2, 0) = F$ if $h(t_1) \neq h(t_2)$.

(3) If t_1 and t_2 are not both additive terms then $I(t_1 = t_2, 0) = N$.

Values for all nonatomic sentences and a corresponding theory for each t are then determined as in Definition 2.9.

Proposition 6.1. I determines a functional theory at all times.

Proof. It is obvious that this is so (relative to the functionality of the classical theory of the real numbers) at times other than $t = 0$, since the quantity space (space of values of f) behaves identically with the real numbers at those times. So suppose time= 0 and let $t_1 = t_2$ hold. That is $I(t_1 = t_2, 0) \in \{B, T\}$; and so by construction of I both t_1 and t_2 are additive. Therefore any atomic sentence Ft_1 is additive iff Ft_2 is additive. As usual it suffices to prove the proposition for only one replacement of t_1 by t_2. If neither Ft_1 nor Ft_2 is additive then neither hold by construction of I. If both are additive then let Ft_1 hold and be of the form $t(t_1) = t_3$, so that Ft_2 is $t(t_2) = t_3$. Now since $t(t_1) = t_3$ holds, $h(t(t_1)) = h(t_3)$. But since $t_1 = t_2$ holds, $h(t_1) = h(t_2)$. So by the fact that $+'$ and $-'$ are functions on $[0, |k_2 - k_1|)$, $h(t(t_2)) = h(t(t_1))$; and since the latter $= h(t_3)$, Ft_2 holds. If Ft_1 does not hold, then $h(t(t_1)) \neq h(t_3)$; whence by $h(t_1) = h(t_2)$, $h(t(t_2)) \neq h(t_3)$, so that Ft_2 does not hold. The argument is similar if Ft_1 is $t_3 = t(t_1)$. Hence I determines a functional theory at time= 0 also. $\square$

Proposition 6.2. The inconsistent zero degree theory determined at $t = 0$ extends the additive part of classical real number theory (R_c).

Proof. By induction on the number of occurrences of $\{\sim, \&, \forall\}$ in an additive sentence A, that (i) if A is true in R_c then $I(A, 0) = T$ and (ii) if A is false in R_c then $I(A, 0) \in \{F, B\}$.

(Base:) Clear by construction of I.

($\sim$clause (i):) If $\sim A$ is true in R_c then A is false in R_c; so by inductive hypothesis (ii) $I(A, 0) \in \{F, B\}$, so $I(\sim A, 0) = T$.

($\sim$clause (ii):) If $\sim A$ is false in R_c then A is true in R_c; so by inductive hypothesis (i) $I(A, 0) = T$, so $I(\sim A, 0) = F \in \{F, B\}$.

(&clause (i):) If $C \& D$ is true in R_c then both C and D are true in R_c so by inductive hypothesis (i) $I(C, 0) = I(D, 0) = T$, so $I(C \& D, 0) = T$.

(&clause (ii):) If $C\&D$ is false in R_c then one or both of C, D is false in R_c. Let C be false. By inductive hypothesis (i) $I(C,0) \in \{F, B\}$, so that $I(C\&D, 0) \in \{F, B\}$.

($\forall$clause (i):) If $(\forall x)Fx$ is true in R_c then Ft_1 is true in R_c for every name t_1. By inductive hypothesis (i) $I(Ft_1) = T$, so that $I((\forall x)Fx, 0) = T$.

($\forall$clause (ii):) If $(\forall x)Fx$ is false in R_c then Ft_1 is false in R_c for some name t_1. By inductive hypothesis (ii) $I(Ft_1, 0) \in \{F, B\}$, so that $I((\forall x)Fx, 0) \in \{F, B\}$. $\square$

Summarising, the theory describing the value space of f and g' is consistent complete R_c at times other than $t = 0$. At $t = 0$, the theory in its additive part is inconsistent, complete and contains R_c. In its nonadditive part it is incomplete, and overall it is functional. The additive part of R_c holds at all times, and at the singularity $t = 0$ the extra additive propositions $0 = k_2 - k_1$, $k_1 = k_2$, $k_1 + 1 = k_2 + 1$, *etc.* hold, making things inconsistent at that time.

Consider any continuous function $g(t)$ with the property that at all but a finite set $\{t_i : 1 \leq i \leq n\}$ of times/points g is differentiable; and that for all the t_i, left and right derivatives are defined (but classically unequal). This determines a closed-set topological space on the real numbers whose boundaries are the singletons $\{t_i\}$ and whose closed sets have the basis $[-\infty, t_1], [t_1, t_2], \ldots, [t_{n-1}, t_n], [t_n, +\infty]$. So one can identify the derivative function g' with a map from the set of times with the closed set topology to (the set of $PJ4$-theories of $\mathcal{L} \times$ the real numbers), with the following two provisos:

(1) if $\{t\}$ is not a boundary then $g'(t) = \langle R_c,$ the classical derivative of g at $t\rangle$;

(2) if $\{t\}$ is a boundary then $g'(t) = \langle Th_t,$ the left hand derivative at $t\rangle$, where Th_t is the additively inconsistent multiplicatively incomplete theory described above. Note that it would be equivalent if the right hand derivative were used instead; since in Th_t, LH derivative = RH derivative. The continuous function g can itself be regarded as such a map, which shows that there is a proper generalisation here. For g, in which there are no boundaries, the closed set topology is just $\{R, \Lambda\}$; and for all t, $g(t) = \langle R_c,$ the classical value of g at $t\rangle$. For any such function g, g is differentiable at t if $g'(t)$ is defined; which it is at every t, so g is differentiable

everywhere and its derivative is g'. Finally, a function is continuous at t if the LH limit at t = the RH limit at t = the value of the function at t. But this holds for g' at all t. Hence g' is continuous and g has a continuous derivative.

It might be wondered whether the derivative $g'(0)$ should contain information about whether g is increasing, decreasing, or stationary at $t = 0$. This would seem to need an absolute distinction between positive and negative values of the derivative at $t = 0$; though since such a distinction could be programmed, no doubt there is functionality at some level. But in any case it isn't clear that there are straightforward intuitions about whether g is increasing if according to its LH derivative it is increasing but according to its RH derivative it is decreasing. (Stationary?) Perhaps the only cases where intuition gives a lead are when g is increasing according to both LH and RH derivatives, or decreasing according to both. But this conception already imports classical LH and RH neighbourhoods of $t = 0$. Thus one might say that the inconsistent account describes the 'magnitude' of the rate of change of g, an 'instantaneous slip' from k_1 to k_2; while the direction of change, where it is meaningful at all, is a matter of neighbourhoods.

A function 'defines its own logic' in that it is associated with a closed set topology which constitutes a paraconsistent logic (see Chapter 11). The propositions of the logic are closed sets of times at which various additive atomic equations hold. The negation of an additive proposition holds at the closure of the set-theoretic complement of times at which the proposition holds, so an additive proposition and its negation hold at the boundary. Multiplicative (nonadditive) propositions, however, hold only on open sets and thus fail to hold at the boundaries, as expected. The composite open-closed-boundary-space is a Boolean algebra with open set (intuitionist) and closed set (paraconsistent) subalgebras or sublogics, which describe the values of multiplicative and additive propositions respectively.

CHAPTER 7: THE DELTA FUNCTION

1. Introduction

In Chapter 6, differentiating the function $g(t) = k_1 t$ for all $t \leq 0$ and $g(t) = k_2 t$ for all $t \geq 0$, where k_1 and k_2 are classically distinct real numbers, led to the inconsistent continuous function $f(t) = g'(t) = k_1$ for all $t \leq 0$ and $f(t) = k_2$ for all $t \geq 0$. If a dynamic system is described by $g(t)$, then the derivative takes an instantaneous jump at $t = 0$. That it is instantaneous rather than taking an infinitesimal amount of time, is represented by the inconsistent continuity. The aim in this chapter is to differentiate one step further, finding $f'(t)$. The new derivative can naturally be thought of as $(k_2 - k_1).\Delta(t)$, where $\Delta(t)$ has the two properties (i) $\Delta(t) = 0$ for all $t \neq 0$, and (ii) $\int_{-\infty}^{+\infty} \Delta(t)dt = 1$. Property (ii) means that a constant times the integral recovers the precise amount of the jump from k_1 to k_2. The delta function occupies an interesting niche in the history of mathematics. Long regarded as problematic but useful in elementary quantum theory and quantum field theory, it was eventually 'solved' in Schwartz' Theory of Distributions; but at the cost of a considerable increase in complexity, as well as an increase in the size of the function space for quantum mechanics.

One can produce an account of something close to the delta function within classical nonstandard analysis. Considering the t-axis (the real line) and its value space as augmented by infinitesimals and their reciprocals the infinite numbers, one can draw a triangle with base 2δ and height $1/\delta$. The area of this triangle is (base$\times$height)$/2 = 1$. The triangle is described by the function $\Delta_1(t) = 0$ for all t with $|t| \geq \delta$, $\Delta_1(t) = (1/\delta^2).t + (1/\delta)$ for $-\delta \leq t \leq 0$, and $\Delta_1(t) = (-1/\delta^2).t + (1/\delta)$ for $0 \leq t \leq \delta$. The slopes of the sides are $1/\delta^2$ for $-\delta \leq t \leq 0$ and $-1/\delta^2$ for $0 \leq t \leq \delta$. Property (i) for the delta function, namely $\Delta(t) = 0$ for all $t \neq 0$, is not quite right here. But it is nearly right, since $\Delta_1(t) = 0$ for all *real* $t \neq 0$. Property (ii) for Δ_1 follows straightforwardly from any reasonable account of the integral as an area function, given the area under the triangle as described. This account of the delta function differs from Robinson's approach in nonstandard analysis, which is more like a nonstandard account of the theory of distributions, in keeping with Robinson's operationist attitude to infinitesimals. But it is a reasonable account all the same.

Nevertheless, it is possible to exploit inconsistency to give an account where $\Delta(t) = 0$ for all nonzero t; and which more reasonably gives an *instantaneous* account of the change in $f(t)$ from k_1 to k_2. Furthermore it is possible to give sense to the idea that the rate of change is different in an instantaneous jump from k_1 to k_2, from an instantaneous jump from k_1 to k_3.

2. Functionality

In Chapter 5 the reciprocals of infinitesimals were avoided because they give problems with functionality. But these problems are not quite so insurmountable. It is shown in this chapter that a construction can be given wherein total functionality fails but some partial and reasonable control of functionality remains. Further, the area of failure of functionality has a reasonable motivation. In turn, this leads to an inconsistent account of the delta function.

First, introduce the concepts of 'the bandwidth of zero' and 'the representative of zero'. In Chapter 5 certain infinitesimals were inconsistently identified with zero, while others were consistently nonzero. One can think of zero as 'smeared out' over the range or bandwidth of the former. A positive infinitesimal δ is taken as the bandwidth of zero, and δ^2 is taken as the representative of zero. The idea is that all and only those numbers infinitesimal w.r.t. δ are inconsistently identified with zero, while δ^2 represents zero for the purposes of multiplication. The bandwidth of zero gives the equivalence relation $\approx\delta$ of Chapter 5.

Definition 7.1. The theory DR is given by

(1)   background logic $RM3$

(2)   names for every hyperreal number finite or infinitesimal w.r.t. $1/\delta^2$, that is names for all numbers x such that $x.\delta^2$ is finite or infinitesimal

(3)   term-forming operators $\{+, -, \times, /\}$

(4.1)   If t is any name then $I(t) = [t]$

(4.2) $I(+)$ and $I(-)$ are addition and subtraction on the set of equivalence classes of numbers finite or infinitesimal w.r.t. $1/\delta^2$

(4.3)   For multiplication, if t_1 and t_2 are terms with $[t_1] = [0]$, then $I(t_1 \times t_2) = I(t_2 \times t_1) = [\text{the representative of zero}.t_2] = [\delta^2.t_2]$. (Note that here $\times$ is the term forming operator and $.$ is multiplication between hyperreal numbers.)

(4.4)   If neither $[t_1] = [0]$ nor $[t_2] = [0]$ then: (4.4.1) if not both ($[t_1]$ is infinite and $[t_2]$ is infinite), then $I(t_1 \times t_2) = [t_1.t_2]$; else (4.4.2) if both $[t_1]$ and $[t_2]$ are infinite, then $I(t_1 \times t_2)$ is not defined.

(5) $I(t_1 = t_2) = T$ if $t_1 = t_2$, $I(t_1 = t_2) = B$ if $t_1 \neq t_2$ but $I(t_1) = I(t_2)$, else $I(t_1 = t_2) = F$.

The notion of infinitude of equivalence classes is well defined in that every member is an infinite hyperreal number. Similarly for middlesized and infinitesimal numbers. Addition and subtraction of numbers finite or infinitesimal w.r.t. $1/\delta^2$ remain noninfinite w.r.t. $1/\delta^2$, so there is no need to restrict these operations. Division and reciprocation are set aside for a while.

Proposition 7.2. (1) For any term t, $\vdash t \times 0 = 0$ iff $t.\delta$ is infinitesimal. (2) If t is noninfinite, then $\vdash t \times 0 = 0$.

Proof. (1) $\vdash t \times 0 = 0$ iff $I(t \times 0 = 0) \in \{B, T\}$ iff $I(t \times 0) = I(0)$. Now $I(t \times 0) = [\delta^2.t]$ and $I(0) = [0]$. But $[\delta^2.t] = [0]$ iff $\delta^2.t$ is infinitesimal w.r.t. δ, that is iff $\delta^2.t/\delta$ is infinitesimal, iff $t.\delta$ is infinitesimal. (2) If t is noninfinite then $t.\delta$ is infinitesimal; then apply (1). □

Thus the use of a representative of zero does not disturb the 'nullifying' property of multiplication by zero of noninfinite numbers (*e.g.* reals). Similarly some but not all infinite numbers with names in the theory are nullified by zero. For example, let H be the name of the infinite number which is the reciprocal of δ^2 in the classical hyperreals, let H_1 be the name of $2/\delta^2$, let H_2 be the name of $1/\delta$, and let H_3 be the name of $1/\delta^{2/3}$. All these names occur in the theory since $H_1\delta^2$ is finite. But $\vdash H \times 0 = 1$ and $\vdash H_1 \times 0 = 2$ and $\vdash H_2 \times 0 = \delta$ and $\vdash H_3 \times 0 = 0$, (though not $\vdash H_2 \times 0 = 0$, since δ/δ is not infinitesimal).

As was seen in Chapter 5, multiplication of noninfinite numbers by noninfinite numbers is functional. But this is not in general true of multiplication of noninfinite numbers by infinite numbers. For example, $\vdash 5 = 5 + \delta^2$ and $\vdash H = H + \delta^2$ but not $\vdash 5 \times H = (5 + \delta^2) \times (H + \delta^2)$ nor even $\vdash 5 \times H = (5 + \delta^2) \times H$. This is because $[(5 + \delta^2).H] = [(5.H) + (\delta^2.H)] = [(5.H) + 1] \neq [5.H]$, so that $I((5 + \delta^2) \times H) \neq I(5 \times H)$. Multiplication is however perfectly well-defined in the theory. The failure of functionality at this point does not seem particularly troublesome or bad. It is related to resolution of the smearing of zero, in that multiplication of the zero difference between two numbers by a sufficiently large infinite number can produce a nonzero (though possibly still infinitesimal) difference. We also have:

Proposition 7.3. If $I(t_1 \times t_2)$ is defined, then it is noninfinite w.r.t. $1/\delta^2$; similarly for addition and subtraction.

Proof. Names are restricted to numbers which are noninfinite w.r.t. $1/\delta^2$. The sum and difference of any such numbers are likewise restricted; while multiplication is defined only when at least one number is finite, and the product of a finite number by one noninfinite w.r.t. $1/\delta^2$ is likewise. $\qquad\square$

Reciprocals and division can be added to the theory, for example by setting $I(t^{-1}) = [t^{-1}]$ for all t such that neither $\delta.t$ nor δ/t is infinite (e.g. all reals, $\delta^{1/2}$, $\delta^{-1/2}$, etc.). This gives $I(t \times t^{-1}) = [t.t^{-1}] = [1]$, thus $\vdash t \times t^{-1} = 1$. However reciprocation is not everywhere functional. For example, let $t_1 = \delta$ and $t_2 = \delta + \delta^2$. Then $\vdash t_1 = t_2$; but $(t_1^{-1} - t_2^{-2})/\delta = t_1^{-1}.t_2^{-1}(t_2 - t_1)/\delta = (\delta^2 + \delta^3)^{-1}.\delta^2/\delta = 1/(\delta + \delta^2)$ which is not infinitesimal, so that not $\vdash t_1^{-1} = t_2^{-2}$. Reciprocation of small enough nonzero numbers thus 'unsmears identity'. As usual, however, reciprocation is functional on the middlesized numbers: if $\vdash t_1 = t_2$ and both are middlesized; then since $(t_1^{-1} - t_2^{-1})/\delta = (t_1 t_2)^{-1}.(t_2 - t_1)/\delta$ and the latter is infinitesimal, so is the former.

Summarising, the theory is functional for $\{+, -\}$; and functional for multiplication where it is desirable, on middlesized numbers. Elsewhere, identities are unsmeared by multiplication by large enough numbers. Multiplication by zero nullifies middlesized numbers, as well as all infinitesimals and some infinite numbers. Multiplication of zero by big enough numbers unsmears the zero to produce a product which is at most middlesized, and which depends on the size of the infinite number. There is no biggest infinite number but there is a biggest order of size, in that numbers infinite w.r.t. $1/\delta^2$ do not exist.

The latter dependency can be exploited in an account of derivatives of inconsistent continuous functions.

Definition 7.4. $\Delta(t)$ is the function from $\{x : x$ is hyperreal and $x.\delta^2$ is noninfinite$\}$ to itself; with the property that $\Delta(t) = 0$ if $t = 0$ does not hold, else $\Delta(t) = H$.

To find the area function for Δ in DR, draw at zero a rectangle of base zero units (that is a vertical line above zero) with height H. Take the infinite Riemann sum over the interval $[a, b]$ including 0 w.r.t. the partition determined by the infinitesimal $dt = \delta^2$. The contributions of all terms for which $\Delta(t_i) = 0$, are obviously zero. The only nonzero contribution to the infinite Riemann sum is the area of the line of height H at $t = 0$. The area of a line of finite height is, naturally enough, zero: base × height $= 0$ × noninfinite number $= 0$. But the area of a line of infinite length can be nonzero: $0 \times H = \delta^2 \times H = 1$. Clearly also, this is the magnitude of a unit instantaneous jump in an inconsistent continuous function. Thus $\Delta(t)$ can serve as the derivative of such functions, and conversely the integral of Δ is the instantaneous one unit jump. Should the instantaneous jump be from

k_1 to k_2, there is then a natural measure of the magnitude and sign of the rate of jump in $(k_2 - k_1).\Delta(0)$, that is $(k_2 - k_1).H$. Note also that the Δ-function can be translated r units along the t-axis by $\Delta_r(t) =df \ \Delta(t+r)$.

This method of integration doesn't work with the area functions of Chapter 5, since there resolution of the t-axis below the order of magnitude of δ^2 is forbidden. So one can say that the normal area functions of CR apply until the function jumps instantaneously, and then its derivative is in a different space DR where Riemann sums are taken w.r.t. partitions of size 0, that is δ^2. Needless to say, it is not claimed that this is the 'right' account of the delta function, only that it is *an* account. Also, it is not apparent how to go on to differentiate delta functions in turn by these methods, but then there appear not to be any meaningful intuitions about that anyway.

CHAPTER 8: INCONSISTENT SYSTEMS OF LINEAR EQUATIONS

1. Introduction

The existence of the inconsistent case of a system of linear equations (or for that matter any system of constraints, not necessarily linear) has been known for a long time, but there has been no attempt to analyse its structure. There would seem to be good reason to do so, if only because the state of affairs might arise in a real life control system (see sections 3 and 4). Using the methods developed so far, it is possible to say something about the structure of solutions to such cases; though it must be confessed that in the end the situation remains less than satisfactory.

2. The Inconsistent Case

Consider a system S of n linear equations in s unknowns $x_1, \ldots, x_s$ having an $n \times s$ coefficient matrix $Mc = [a_{ij}]$ and an $n \times (s+1)$ augmented matrix $Ma = [Mc, B]$, where $B = \text{col}[b_1 \ldots, b_n]$ is the column vector of constants. The usual story is that S has a solution iff the row rank r of $Mc =$ the row rank of Ma, and S has a unique solution iff in addition $r = s$. We concentrate mostly on the first of these, looking briefly at the second later. Clearly $r \leq \text{rowrank}(Ma)$, since every row of Mc is part of a row of Ma. But if $r < \text{rowrank}(Ma)$, then elementary row operations on Ma will produce an equivalent matrix with zeros everywhere below row r except in column $s+1$. This corresponds to the equations $0 = b_{r+1}, 0 = b_{r+2}, \ldots, 0 = b_n$ where one or more of the b_{r+i} are nonzero. This is an inconsistency, so that it is impossible to satisfy S below row r. Hence S has no solutions. (See *e.g.* Perlis [44], Birkhoff and MacLane [4].)

But if there were inconsistent arithmetics in which $0 = b_{r+1}, 0 = b_{r+2}, \ldots$ could all hold, then there would be no particular reason why all of the equations of S could not hold simultaneously. So we can begin by informally postulating structures in which $0 = b_{r+1}, 0 = b_{r+2}, \ldots$ hold inconsistently, that is where $\sim 0 = b_{r+1}$, $\sim 0 = b_{r+2}, \ldots$ and a reasonable amount of classical real number theory also hold.

The distinctive role played in S by the column vector of constants motivates the following definition, particularly part (3).

Definition 8.1. (1) A matrix M is *row reduced* if (a) every leading entry of a nonzero row is 1, and (b) every column containing such a leading entry 1 has all other entries zero.

(2) M is in *row echelon form* (REF) if also (c) each zero row comes below all nonzero rows, and (d) leading coefficients begin further to the right as one goes down.

(3) An augmented matrix $Ma = [Mc, B]$ is in *weak row echelon form* (WREF) if Mc is in REF.

Thus Ma in WREF might look like

$Ma1$:

$$\begin{bmatrix} 1 & 0 & 0 & 2 & 0 & 3 \\ 0 & 1 & 0 & 0 & 0 & 2 \\ 0 & 0 & 1 & 2 & 3 & 2 \\ 0 & 0 & 0 & 0 & 0 & 5 \\ 0 & 0 & 0 & 0 & 0 & 7 \end{bmatrix}$$

when deleting the last column produces a matrix in row echelon form.

It is obvious that any matrix regarded as an augmented matrix can be transformed to one in WREF using elementary row operations, and that some sequence of operations on an augmented matrix Ma suffices to reduce its coefficient matrix Mc to its unique REF. Further, Ma arises from a consistent set of equations just in case there are no rows with a nonzero only in the last column. Note that the usual process of determining a basis for the rowspace of Ma considered simply as a matrix (rather than as an augmented matrix arising from a set of equations) goes beyond WREF to produce REF in Ma, that is a 1 in row $r + 1$ (where rowrank(Mc) = r) and zeros everywhere else in that column. But as we see this is not so useful for dealing even with the consistent case, so we deal with WREF's. These diverge from REF's only when $\mathrm{rowrank}(Mc) < \mathrm{rowrank}(Ma)$, that is when S is an inconsistent set of equations.

The rowrank of Mc can be read off from any WREF row-equivalent to Ma, as the number of nonzero rows discounting the last place. The rowspace of any matrix in WREF is identical with that of its unique REF. It does not seem to, be determinate whether one should say that the rowrank of Ma is the number of nonzero rows in the WREF, or the generally lesser number of $\mathrm{rowrank}(Mc) + 1$. The latter is favored by the classical treatment of linear algebra, but the former has certain advantages in inconsistent situations.

In general more than one sequence of elementary row operations suffices to reduce Mc to REF. If and only if S is consistent, exactly those sequences of row operations reduce Ma to REF. If S is inconsistent, just those sequences of row operations reduce Ma to WREF. Furthermore, each such sequence produces a unique vector of constants in column $s + 1$ with at least one nonzero entry below row r. However, the rowspace spanned by the first r row vectors of a WREF of Ma is not unique. For example, consider
$Ma2$:

$$\begin{bmatrix} 1 & 0 & 0 & 2 & 0 & 3 \\ 0 & 1 & 0 & 0 & 0 & 2 \\ 0 & 0 & 1 & 2 & 3 & 2 \\ 1 & 0 & 0 & 2 & 0 & 8 \\ 0 & 1 & 0 & 0 & 0 & 9 \end{bmatrix}$$

Subtracting row 1 from row 4 and row 2 from row 5 gives the WREF
$Ma3$:

$$\begin{bmatrix} 1 & 0 & 0 & 2 & 0 & 3 \\ 0 & 1 & 0 & 0 & 0 & 2 \\ 0 & 0 & 1 & 2 & 3 & 2 \\ 0 & 0 & 0 & 0 & 0 & 5 \\ 0 & 0 & 0 & 0 & 0 & 7 \end{bmatrix}$$

While first interchanging the first and fourth rows then performing the same operations gives the WREF

$Ma4$:

$$\begin{bmatrix} 1 & 0 & 0 & 2 & 0 & 8 \\ 0 & 1 & 0 & 0 & 0 & 2 \\ 0 & 0 & 1 & 2 & 3 & 2 \\ 0 & 0 & 0 & 0 & 0 & -5 \\ 0 & 0 & 0 & 0 & 0 & 7 \end{bmatrix}$$

where the rowspace spanned by the first three rowvectors is different in $Ma3$ from $Ma4$. Nonetheless, two aspects are determinate.

(a) Given a selection of r linearly independent rowvectors, reducing just these to REF uniquely determines the constants $b_{r+1}, b_{r+2}, \ldots$ lower down (since in the coefficient matrix Mc every row lower down than row r is a unique linear combination of earlier rows, and this linear combination carries the b_{r+i} along with it).

(b) There are only a bounded number of selections of r linearly independent rowvectors from the $n \times (s+1)$ matrix Ma.

So we can think of an augmented matrix reduced to WREF as containing two parts: (i) the first r rows, spanning a (classical) r-dimensional vector subspace of $V_{s+1}(F)$, and constituting a consistent set of linear relationships between the variables $x_1 \ldots x_s$ which can be satisfied in some classical $(s-r)$-dimensional solution space (which is exactly how one would say it in the case where $r = \text{rowrank}(Ma)$, see [4], p.169); plus (ii) a set of propositions of the form $0 = b_{r+1}, 0 = b_{r+2}, \ldots$ to be satisfied conjointly in an inconsistent arithmetic. (It is also desirable that these parts interact.) And furthermore there are a bounded number of consistent solution spaces and each determines a unique set of inconsistent arithmetical propositions.

So one can informally define a solution structure for a set S of n linear equations in s unknowns to be a finite collection $\{C_j\}$ where each C_j has two interacting parts, a classical $(s-r)$-dimensional solution space, and a structure in which a set of inconsistent identities $0 = b_{r+i}$ hold in an appropriate paraconsistent logic. A solution, then, would be a vector of constant values for $x_1 \ldots x_s$ satisfying the set of relationships of the consistent solution space (that is, lying in one of the $(s - r)$-dimensional classical solution spaces), and interacting with the corresponding inconsistent mathematical theory. In the case where S is consistent, these reduce to the classical definitions of solution space and solution, since the second part of the definitions covering inconsistency becomes inoperative.

The situation would be less flexible if it was permitted to go beyond WREF to REF for Ma by reducing all or even any of the $0 = b_{r+i}$ to $0 = 1$ by multiplying by $(b_{r+i})^{-1}$; and even less flexible if one then cleared all other places in $\mathrm{col}[b_1 \ldots b_n]$. Aside from mathematical interest, at least one good reason for disallowing this is a functional one. Division by zero remains functionally chaotic despite all attempts to do so inconsistently. But if division by zero is disallowed, then division by anything equal to zero, such as $b_{r+1}, \ldots$, ought also to be disallowed, on pain of failure of functionality. That is, when $0 = b_{r+1}$ holds then $(b_{r+1})^{-1}$ should be undefined. But it is $(b_{r+1})^{-1}$ which is needed to be the multiplier to reduce $0 = b_{r+1}$ to $0 = 1$, or to $0 = b_{r+2}$ for that matter. That is to say, when dealing with inconsistent sets of equations, reduction to WREF is correct and preferred to REF, while there is no disagreement in the consistent case. A slight problem arises because in arriving at a WREF, multiplication by $(b_{r+1})^{-1}$ might have been used in reducing Mc to REF and it might seem that even this move should be disallowed. Against this it seems fair to say that in reducing Mc to REF one is remaining within the consistent part of the solution space, so one is entitled to use $(b_{r+1})^{-1}$ with its classical meaning.

3. Control Theory

Modern control theory describes control systems in terms of a (column) vector u of inputs, a vector x of inner states, and a vector y of outputs. A plant functioning stably can be described as a linear transformation (matrix) M representing the laws governing the plant, and transforming input into output in accordance with $y(t) = M.u(t)$, where t is the time variable. A more detailed analysis of such plants, incorporating feedback and the state vector x, is standardly given by supposing four matrices A, B, C, D with the two relations: $x(t+1) = Ax(t) + Bu(t)$ and $y(t) = Cx(t) + Du(t)$. However, it is not necessary to incorporate those relations at the present stage.

An unexpected and persistent change is postulated in the output. Since the change is significant and nontransient, it can be regarded as originating from a change in the physical laws of the plant described by M. This leads to a distinction between the matrices Mold and Mnew. Mold (the original M) is responsible for the predicted value of the output, $y$pred; and Mnew is responsible for the observed value of the output, $y$obs. Mold was known when the plant was functioning correctly; Mnew is unknown but its result $y$obs is known through observation. One can now define a machine to be *malfunctioning* iff $y$obs $\neq y$pred, otherwise *wellfunctioning*.

A standard situation in modern control theory is to determine from an observed stream of outputs $y$obs(t) what is the nature of the linear plant, Mnew, responsible for them. The approach taken here is different. There might be a real-time problem: the problem of keeping some control before Mnew is ascertained, let alone rectified back to Mold. So one wants to see if there are ways of operating the plant (modifying the stream of inputs) under the anomalous conditions without complete shutdown or explosion.

Applying ideas from earlier in this chapter, one would like to form a model of the malfunctioning plant which exploits inconsistency to represent quantitatively the discrepancy between $y$pred and $y$obs. A desirable constraint would be that the inconsistency disappears when the plant wellfunctions. Accordingly, one can form the *augmented checkmatrix* for a plant, consisting of a core which is Mold, an extra column which is $y$obs, and a bottom row (checkrow) which is the sums of each of the columns except that the RH cell is Σypred. Subjected to $w$ref, it can be shown that the RH corner entry is zero if $y$obs $= y$pred. When the corner entry is nonzero, it is a parameter which identifies an inconsistent mathematical environment which can be represented in a software controller. There are a number of options for the way in which input can be modified by the controller with the aim of eventual wellfunctioning.

That is, the controller forms an *inconsistent model* of the nature of the malfunctioning plant. A desirable behaviour for malfunctioning plants is that, under the influence of the inconsistent controller, the plant eventually becomes wellfunctioning simply by means of modifying the input. This proves to be possible. On the other hand, among plants which are *not* eventually wellfunctioning there can be defined several types.

(a) A plant *cycles* iff $yobs(t) = yobs(t + k)$ for some k and all t (or all t after some appropriate t_0).

(b) A plant is *persistent* iff $yobs(t) = yobs(t + k)$ for all k and all t (after some appropriate t_0). Note that it is possible to have a malfunctioning but persistent plant, i.e. in which $yobs(t)$ remains constant but never equals $ypred(t)$.

(c) A plant is *bounded* iff no component of $yobs(t)$ ever gets more than a fixed number k from zero, for all t. A plant may be bounded without either cycling or persistence.

(d) A plant is defined (operationally) to *explode* iff some component of $yobs$ exceeds a predetermined bound (in software simulations it has been taken to be 10000).

For a plant which does not eventually wellfunction, any of the above behaviours (a)–(c) is a substitute in which functioning is not too degraded, so all of these behaviours are more desirable than explosion.

All of the above behaviours have been observed in software simulations.

4. Applications, Problems and Special Cases

An application might be a machine with a range of sensor inputs, controlled by a system of simultaneous equations. Supposing that one sensor becomes faulty, and begins returning a value zero for a particular variable, the matrix might go inconsistent and become impossible to invert to find stable settings. In the circumstance one does not want chaos, and it might be that the best fault-tolerant software is one which rides with the contradiction until things are straightened out.

There is no problem about the logical aspects of solution structures for inconsistent systems of simultaneous equations, as is clear from previous chapters. But functionality is more than usually problematic. The difficulty is to find a theory in which all the b_{r+i} are identified with zero and their inverses undefined. The obvious manoeuvre is to go to an additive group as in Chapter 6, undefining all multiplication but integer multiplication. The worst case here would be trying to satisfy jointly, say, $0 = \pi$ and $0 = 2$, since there is no greatest common integer divisor for a single base equation. Unfortunately, the worst case looks to be the typical case. Nevertheless, there are a number of special cases which are more tractable.

(1)   If the field F over which the vector space is constructed is the integers modulo a finite or infinite prime, then there is no problem in taking the greatest common divisor of the b_{r+i}. From the base $0 = b_{r+i}$ all the others follow, and the vector or vectors in the solution space are readily represented in the inconsistent arithmetic. Needless to say if the b_{r+i} are relatively prime then the base is $0 = 1$, which is bad if one remains with integers. But one can be lucky and avoid total collapse of functional structure when they are not relatively prime, which might serve to avoid total degeneration of information in a control system. (Birkhoff and MacLane treat the consistent case, see [4], pp.40-44.)

(2) If the field is R, and there is only *one* $0 = b_{r+i}$, and in the additive group on the interval $[0, b_{r+i})$ the vector relationships of the first r rows of the WREF can be represented, then combined with a judicious change of scale to expand that interval, this might be a practical proposal for control systems. Also, it appears fairly straightforward to separate several inconsistencies into several 'logical dimensions' or 'logical subspaces' (with logical projection operators) in each of which only one inconsistency holds. This option is available and useful whatever one does to superpose the inconsistent substates. If such logical separation could correspond, or be made to correspond, to relative causal separation of subparts of the control system, then reasonable control might be achieved for each of the separately inconsistent subsystems, or at least the area of lack of control isolated (always a paraconsistentist ideal).

(3) The role of computers as engines of empirical arithmetic can't be overlooked. Supposing that it is decided to input all data in the data-type *integer* (which is by no means impossible if a judicious choice of scale size is made), then for example Pascal comes with built-in maximum and minimum integers, just as in the finite inconsistent case. (e.g. MaxInt $= 32767, -$MaxInt $= -32768$, MaxLongInt $= 2147483647$.) The inconvenience of small numbers compared with floating-point numbers might be outweighed by increased control over inconsistent situations. There might be no reason to switch on the fault-tolerance module until inconsistency manifests itself. But even floating-point arithmetic is finite (and inevitably 'approximate', that is approximate to something called a real number). Empirical arithmetic exists, and all computer calculations are carried out in it. So there does not seem to be anything wrong in principle with integer representations of a problem. (If there is a difficulty it is with how finite human brains can imagine that finite output represents something infinite, a real number.) But if integer representations are always OK for a control system, then perhaps there is no real difficulty in always exploiting their inconsistency-tolerance.

(4) A different style of solution is to associate the rows below r with inconsistencies in different dimensions in the state space of the plant: $0 = b_{r+1}$ gm, $0 = b_{r+2}$ volts, ... and the like. Thus inconsistency in one dimension of the phase space ought not to affect consistency or inconsistency in an orthogonal dimension. This approach is still in an early stage of development.

CHAPTER 9: PROJECTIVE SPACES

1. Introduction

A common construction (following more-or less that of Birkhoff and MacLane [4]) of the projective plane $P_2(F)$ over a field F, begins with the 3D vector space $V_3(F)$ and then 'identifies' nonzero triples of members of F, $(x_1, x_2, x_3) = (x_1', x_2', x_3')$ if for some nonzero $a \in F$ and for all $i \leq 3$, $x_i = ax_i'$. The equivalence classes so formed are the points of $P_2(F)$. The lines of $P_2(F)$ correspond to, or 'are', planes in $V_3(F)$; that is *sets* of solutions (x_1, x_2, x_3) of linear homogeneous equations $\sum a_i x_i = 0$ where each $a_i \in F$ and not all $a_i = 0$. One can then identify lines by (or 'identify' lines 'with') the *triples of coefficients* (a_1, a_2, a_3); noting that if $a \neq 0$, then $a \sum a_i x_i = 0$ determines the same set of solutions as $\sum a_i x_i = 0$, so that the projective line whose coefficients are (a_1, a_2, a_3) is identical with the line whose coefficients are $a(a_1, a_2, a_3)$.

In order to keep track of the source of the triples (x_1, x_2, x_3) in this construction, the notion of *homogeneous coordinates* is introduced. Again the terminology is somewhat anomalous: Birkhoff and MacLane (p.275) for example speak of homogeneous coordinates of points as the triples (x_1, x_2, x_3) 'with the identification' $(x_1, x_2, x_3) = a(x_1, x_2, x_3)$ when $a \neq 0$. With the same 'identification', triples (a_1, a_2, a_3) are homogeneous coordinates of lines. One can then proceed to define '(x_1, x_2, x_3) on (a_1, a_2, a_3)' to mean $\sum a_i x_i = 0$; '(x_1, x_2, x_3) is a point' or '$P(x_1, x_2, x_3)$ to mean '$(\exists a_1, a_2, a_3)((x_1, x_2, x_3)$ on $(a_1, a_2, a_3))$'; '(a_1, a_2, a_3) is a line' or '$L(x_1, x_2, x_3)$' to mean '$(\exists a_1, a_2, a_3)((x_1, x_2, x_3)$ on $(a_1, a_2, a_3))$'; and '(a_1, a_2, a_3) contains (x_1, x_2, x_3)' to mean '(x_1, x_2, x_3) on (a_1, a_2, a_3)'. The familiar duality is easily seen to follow, whereby the transformation $(=, P, L, \text{on}, \text{contains}) \rightarrow (=, L, P, \text{contains}, \text{on})$ preserves true sentences.

However, if (x_1, x_2, x_3) and (ax_1, ax_2, ax_3) are distinct triples then they are not *literally* identical homogeneous coordinates. One should speak rather of *different* co-ordinates of the *same* point, with the recognition that *qua* representations of the one thing, a point, they behave congruently w.r.t. its properties. Similarly, triples (x_1, x_2, x_3) and (ax_1, ax_2, ax_3) are not *literally* identical if they are *distinct* members of the one equivalence class, a point. There should then be an explanation of this common but *literally false* terminology; and of course one which does not disturb the entirely correct orthodox account in terms of equivalence classes just given, but rather augments it. In this chapter an explanation is offered in terms of *literal inconsistent identities*.

First the case of inconsistent vector spaces is considered. In section 3 inconsistent projective planes are constructed, in which homogeneous coordinates are inconsistently identified. It turns out to be easier to inconsistentise projective spaces than vector spaces, as the example of the projective plane over R shows. In section 4, projective planes over consistent and inconsistent fields modulo an infinite prime are constructed. In these the usual projective duality holds for a strengthened set of concepts which contains 'finite' and 'infinite' as well.

2. Vector Spaces

Take background logic $RM3$, and consider any one of the finite inconsistent fields modulo a prime p with names for all the integers, of Chapter 2. Call it F, and construct the theory $V_3(F)$ as follows. Add as terms all triples (t_1, t_2, t_3) where the t_i are terms from F. Take term-forming operators $+$ and . (the latter usually suppressed) for vector addition and scalar multiplication. The domain D has all triples from $\{0, 1, \ldots, p-1\}$, and the operations of vector addition and scalar multiplication on the usual classical vector space constructed on the finite classical field modulo p. Set $I((t_1, t_2, t_3)) = (t_1 \bmod p, t_2 \bmod p, t_3 \bmod p)$; set $I(+) =$classical vector addition mod p and set $I(.) =$classical scalar multiplication mod p. This induces as in Chapter 2,

$$I((t_1, t_2, t_3) + (t_4, t_5, t_6)) = I(+)(I((t_1, t_2, t_3)), I((t_4, t_5, t_6)))$$

and $\quad I(t.(t_1, t_2, t_3)) = I(.)(I(t), I((t_1, t_2, t_3))).$

Finally set $I((t_1, t_2, t_3) = (t_4, t_5, t_6)) = B$ if $I((t_1, t_2, t_3)) = I((t_4, t_5, t_6))$, else $I = F$.

Now, for example, $\vdash (0,1,2) = (p, p+1, p+2)$ & $\sim (0,1,2) = (p, p+1, p+2)$, as well as $\vdash (0,1,2) = (p+1).(0,1,2)$ & $\sim (0,1,2) = (p+1).(0,1,2)$, and $\vdash (0,1,2) + (0,1,0) = (0,2,2)$ & $\sim(0,1,2) + (0,1,0) = (0,2,2)$, and $\vdash \sim(0,1,2) = (0,2,2)$, but not $\vdash (0,1,2) = (0,2,2)$.

Proposition 9.1. $V_3(F)$ is transparent.

Proof left to the reader.

Proposition 9.2. All first order extensional sentences of the classical theory of vector spaces hold in $V_3(F)$.

Proof. Consider the classical finite vector space mod p; that is the above structure modified so that there are only a finite number of names $\{0, 1, \ldots, p-1\}$, and $I((t_1, t_2, t_3) = (t_4, t_5, t_6)) = T$ if $I((t_1, t_2, t_3)) = I((t_4, t_5, t_6))$, else $I = F$. As is well known this satisfies all the axioms for vector spaces and their classical consequences. Add the infinite number of names $\{p + 1, \ldots\}$ and all complex terms generated by the term-forming operators. For any new term t or (t_1, t_2, t_3) set $I(t) = t \bmod p$ or $I((t_1, t_2, t_3)) = (t_1 \bmod p, t_2 \bmod p, t_3 \bmod p)$ respectively. Obviously the addition of the extra names makes no difference to the value of the sentence not containing them, since they map congruently onto existing names (or appeal to Proposition 2.14). Change the model so that $I'(\text{term}_1 = \text{term}_2) = B$ iff $I'(\text{term}_1) = I'(\text{term}_2)$. By Extendability (Proposition 2.10), I' extends I. But I' determines $V_3(F)$, so all extensional sentences of classical vector space theory hold in V_3F. $\qquad\square$

It is also obvious that $V_3(F)$ extends the classical theory of the integers. In addition it is clear that this method of construction of inconsistent vector spaces suffers the same limitations as inconsistent fields, because any limitations on the functionality of the field F translate immediately into limitations on vector addition and scalar multiplication.

3. Projective Geometry

Proceeding to projective geometries over these structures, add to the domain D the equivalence classes $[(x, y, z)]$ of triples from $\{0, 1, \ldots, p - 1\}$, where $[(x, y, z)] = [w(x, y, z)] = [(wx, wy, wz)]$ for all $w \neq 0$, the last three products being mod p. These equivalence classes form a classical projective geometry in the usual way. This can be described in the classical structure $\langle D, I \rangle$ where $I((t_1, t_2, t_3)) = [(I(t_1), I(t_2), I(t_3))]$; and $I(\text{term}_1 = \text{term}_2) = T$ iff $I(\text{term}_1) = I(\text{term}_2)$ else $I = F$; and with 'point', 'line', 'on', and 'contains' defined in the usual way (see above this section, for a more full description see below this section). Notice that in the *language* only the *homogeneous coordinates* (t_1, t_2, t_3) appear as terms, the equivalence classes $[(t_1, t_2, t_3)]$ do not appear. This is where the distinction between objects of the space and their names is being made: coordinate systems are naming systems, a natural view from elsewhere *e.g.* differential geometry. (Nonetheless it is appropriate to take the long term view that this distinction should be blurred, so as to be dealing with mathematical structures and objects with inconsistent properties.)

However this I fails to reflect the origin of the t_i as natural number terms and the (t_1, t_2, t_3) as being inconsistently identified with certain (t_4, t_5, t_6) in the underlying vector space. In addition, the aim of representing homogeneous coordinates in terms of inconsistent identifications is not realised. There are (at least) two structures in which these aims can be more adequately realised.

For both of these structures, it is useful to introduce two new sets of metalinguistic variables $x, x_1, x_2, \ldots$ and $a, a_1, a_2, \ldots$ ranging over names and $\{+, -, \times, /\}$ compositions of them, to reflect the differing roles of points and lines in the projective geometry. If x_1, x_2, x_3 are any such terms, introduce '$P(x_1, x_2, x_3)$' intended to mean '(x_1, x_2, x_3) is a point'. Also, if a_1, a_2, a_3 are any such terms, introduce the predicate '$L(a_1, a_2, a_3)$' intended to mean '(a_1, a_2, a_3) are the coefficients of a line'; and the two relations '(x_1, x_2, x_3) on (a_1, a_2, a_3)' for 'The point (x_1, x_2, x_3) is on the line whose coefficients are (a_1, a_2, a_3)', and '(a_1, a_2, a_3) contains (x_1, x_2, x_3)' for 'The line whose coefficients are (a_1, a_2, a_3) contains the point (x_1, x_2, x_3)'.

These predicates and relations are still only syntax, so it is necessary to specify the interpretation function for sentences containing them, and for identity sentences as well. Only then can the resultant theory be judged as a projective geometry. As I just said there are two ways to do this. The first way is as follows.

Set $I(P(x_1, x_2, x_3)) = B$ if $[(I(x_1), I(x_2), I(x_3))]$ is a point in classical projective geometry on the domain else $I = F$; set $I(L(a_1, a_2, a_3)) = B$ if $[(I(a_1), I(a_2), I(a_3))]$ is the coefficient of a line in the domain; set $I((x_1, x_2, x_3)$ on $(a_1, a_2, a_3)) = B$ if $I((x_1, x_2, x_3))$ is a point and $I((a_1, a_2, a_3))$ is a line and $\sum a_i x_i = 0$; and set $I((a_1, a_2, a_3)$ contains $(x_1, x_2, x_3)) = I((x_1, x_2, x_3)$ on $(a_1, a_2, a_3))$. And then for identity sentences set $I(\text{term}_1 = \text{term}_2) = B$ if $I(\text{term}_1) = I(\text{term}_2)$, else $I = F$. (The specification is somewhat roundabout for the case of finite geometries, but it is adequate and has the merit of permitting easy generalisations where the domain is not finite.)

This model produces an inconsistent projective geometry (see below); but a second structure, which confines the inconsistency in a more sensitive fashion, is as follows. Let the interpretation of 'P', 'L', 'on', and 'contains' be classical, that is change B to T in the interpretation of these predicates. This obviously ensures that the first order sentences in the $\{P, L, \text{on}, \text{contains}\}$ language which hold are exactly those of classical finite projective geometry mod p, including the characteristic $(x, y)((Lx \& Ly) \supset (\exists z)(Pz \& z \text{ on } x \& z \text{ on } y))$, i.e. any lines have a point in common. (It needs identity to state 'exactly one'.) But also continue with $I(\text{term}_1 = \text{term}_2) = B$ if $I(\text{term}_1) = I(\text{term}_2)$, else $I = F$. This has the effect of confining the inconsistency to identity statements between what from the traditional point of view are homogeneous coordinates, and it is only from that source that inconsistency arises. This then is the promised account of 'identifying' homogeneous coordinates: *homogeneous coordinates are inconsistently identified and disidentified just when they are coordinates of the same classical equivalence class.*

This structure is a projective geometry in the sense that all extensional first order sentences of projective geometry hold (by Extendability). There seems to be some difference as to what is the mark of a projective geometry, or what constitute the propositions of projective geometry. Birkhoff and MacLane ([4], p.275) have a simple account: any two distinct points are on a unique line plus duality, both of which hold in the inconsistent structures. Coxeter's ([9], p.230-1) is less simple, but his axioms are still first order, so must hold in the first order classical structures, so in the inconsistent structures. Both inconsistent models above are transparent (proof omitted). Summarising,

Proposition 9.3. There are inconsistent transparent theories in which all sentences holding in classical projective geometry hold.

This has been done here for the finite projective geometries, but it can also be done for $P_2(R)$, the projective plane over R. For all triples of real numbers x_1, x_2, x_3 take the equivalence classes $[(x_1, x_2, x_3)] = [(wx_1, wx_2, wx_3)]$ and set $I(x_1, x_2, x_3) = [(x_1, x_2, x_3)]$ and then give the same conditions on I as above. The point is that this does not collapse from functionality limitations on inconsistent R, because the functional language $\{+, -, \times, /\}$ is discarded in moving to the projective language $\{=, P, L, \text{on}, \text{contains}\}$. This example is discussed further in Chapter 10 in connection with its topological aspects. It is not necessary to have an inconsistent field to start with. The example shows this, but also note that inconsistency in the base field is irrelevant to the construction of the finite geometries above.

4. Projective Geometry Modulo Infinity

This section uses the type of construction of the previous section except that consistent and inconsistent fields modulo an infinite prime are used. In these structures the usual projective duality results can be strengthened to a language containing 'finite' and 'infinite' when these are suitably defined. Sylvester's Theorem is also considered.

There are both consistent and inconsistent projective structures over the fields modulo an infinite prime p. For both, for any terms t_1, t_2, t_3 let $I((t_1, t_2, t_3)) = [t_1 \bmod p, t_2 \bmod p, t_3 \bmod p)]$, where the equivalence classes are equal in the domain just when their members are a multiple mod p of one another. The consistent case is the same as for modulo finite p, with $I(P(x_1, x_2, x_3)) = T$ if (x_1, x_2, x_3) is a point in the domain projective space, else $I = F$; etc.; and $I(\text{term}_1 = \text{term}_2) = T$ if $I(\text{term}_1) = I(\text{term}_2)$, else $I = F$. The two inconsistent cases change T to B respectively (a) for all atomic sentences, or (b) for all identities only. Both are transparent, and (b) confines the inconsistency to the effects of the inconsistent identification of homogeneous co-ordinates as before. Now some definitions are needed.

Definition 9.4. In either consistent or inconsistent theories, a point is *infinite* if it is identical only with points having at least one component x_i which is an infinite nonstandard number, else *finite*. Similarly for lines.

An infinite line is not the same thing as a line at infinity, which does not appear in the present account. Problem: which inconsistent identities are needed to get it?

Proposition 9.5. There are finite points and lines, and infinite points and lines.

Proof. For finite points and lines, consider the equivalence class of any triple all of whose components are finite. For infinite triples, consider $(1, -1, x)$ (any x), that is $(1, p - 1, x)$. If $(1, -1, x)$ were finite, then for some finite t_1, t_2, t_3 $\vdash (1, -1, x) = (t_1, t_2, t_3)$. That is, for some $k \leq p - 1$, $[k(1, -1, x)] = [(t_1, t_2, t_3)]$. So in particular $k. - 1 = t_2$ which is a finite number. Hence k must equal $-t_2$. But it was seen in Chapter 3 that the additive inverse of a finite number is an infinite number (in fact $p - t_2$), so k is an infinite number. But then the first component $k.1$ is equal to t_1 which is thus infinite, contrary to the supposition that t_1 is finite. Hence $(1, -1, x)$ is infinite, either as a point or as the coefficients of a line. $\square$

We recall that the first order duality principle for projective geometry is as follows: if S is any sentence in the language $\{=, P, L, \text{on}, \text{contains}\}$ and S' is the result of interchanging 'P' with 'L' and 'on' with 'contains', then S holds iff S' holds. In the present consistent and inconsistent models there is an extended duality principle, that the same interchange preserves 'holds' in a strong language also containing 'finite' (and *a fortiori* 'infinite'). This might be described as an invariance theorem for 'finite' and 'infinite'. For this, the word 'finite' has to be included in the object language with an appropriate semantic interpretation, so for any number terms t_1, t_2, t_3 whose moduli are not all zero, set $I(\text{Finite}(t_1, t_2, t_3)) = T$ if (t_1, t_2, t_3) is finite, else $I = F$. This obviously induces an interpretation on 'infinite' $=df$ '$\sim$ finite'. Let S be a sentence in the language with 'finite' added, and let S' be the result of interchanging 'P' with 'L' and 'on' with 'contains'.

Proposition 9.6. (Duality) S holds iff S' holds.

Proof. By induction on the number of occurrences of $\{\sim, \&, \forall\}$ to prove that $I(S) = I(S')$.

(Base:) Atomic sentences are of the form $P(\text{term})$, $L(\text{term})$, $\text{Finite}(\text{term})$, term_1 on term_2, term_1 contains term_2, and $\text{term}_1 = \text{term}_2$. But by inspection of I, in all the models $I(P(\text{term})) = I(L(\text{term}))$, and $I(\text{term}_1 \text{ on } \text{term}_2) = I(\text{term}_1$ contains $\text{term}_2)$. Nor do these reversals have any effect on identity or finitude: $I(\text{term}_1 = \text{term}_2)$ holds independently of whether they are points or lines, and $I(\text{Finite}(\text{term}))$ is similarly independent.

($\sim$ clause:) If $I(S) = I(S')$ then $I(\sim S) = I(\sim S')$.

(& clause:) Similar.

($\forall$ clause:) If $I(St) = I(S't)$ for all terms t then $I((x)Sx) = I((x)S'x)$. $\square$

This proposition can be applied after the next one. Let (a_1, a_2, a_3) be the coefficients of a line L where all of $I(a_i) \neq 0 \bmod p$.

Proposition 9.7. If L is finite then every point on L is infinite.

Proof. If L is a finite line and P were a finite point on it then for some (a_1, a_2, a_3) and (x_1, x_2, x_3) where all the a_i and x_i are finite, none of the a_i is classically identical with $0 \bmod p$, at least one of the x_i is not classically identical with $0 \bmod p$, and $\sum a_i x_i = 0$. But this is impossible since the sum of products would be a finite positive number. □

Proposition 9.8. If P is a finite point satisfying the same conditions as L in the previous proposition, any line containing P is infinite.

Proof. By applying duality. □

Note that these break down for any of the following triples/lines/points: $(x_1, 0, 0)$ on $(0, a_2, a_3)$ and $(0, a_2, 0)$ and $(0, 0, a_3)$; $(0, x_2, 0)$ on $(a_1, 0, a_3)$ and $(a_1, 0, 0)$ and $(0, 0, a_3)$ etc. Say that a pair of points *determine* a line if both are on it, and that a pair of lines *intersect* in a point if both contain it. Then,

Proposition 9.9. Every pair of finite points determine an infinite line; every pair of finite lines intersect in an infinite point.

Proof. The first part follows from the previous proposition; the second part follows by duality. □

While a finite point can only be on infinite lines, an infinite point can be on finite lines. Can an infinite point be on both finite and infinite lines? Can an infinite line contain only infinite points?

Finally in this section we consider Sylvester's Theorem. This says that for any positive integer n, if n points are not collinear, then there exists a line through exactly two of them. Sylvester's Theorem holds in classical Euclidean geometry. But it is known that it fails in the classical finite projective planes modulo p. (Reason: In mod p, let $n = p^2 + p + 1$. These $p^2 + p + 1$ points are not all collinear; else some line has $p^2 + p + 1$ points on it, whereas lines in this geometry have only $p + 1$ points on them. But there is no line containing exactly two of them; since again every line has $p+1$ points on it, while the $p^2 + p + 1$ points exhaust the whole space.) But this argument breaks down if p is an infinite prime, provided that n is restricted to *finite* positive integers, because then n cannot be chosen to be $p^2 + p + 1$. So the question is whether Sylvester's Theorem holds for either the consistent or inconsistent projective geometries modulo infinite p? However, if n is allowed to be unrestricted, that is any nonstandard integer as well, then the above argument goes through and Sylvester's Theorem breaks down. Of course this nonstandard version of Sylvester's Theorem (for any finite or infinite n collinear points ...) is a stronger statement, and thus it is not so surprising that it breaks down.

CHAPTER 10: TOPOLOGY

Quotient constructions are a natural place to find inconsistent structures. In this chapter, aspects of the quotient construction in topology are considered. This requires the introduction of the primitive binary predicate $\in$ into the object language. It is seen that there is an interaction between the topological properties of the space from which the quotient topology arises, and the functionality of a natural class of inconsistent models and theories associated with them. A special case of quotient constructions is the ubiquitous practice in topology of joining, cutting or pasting, coming under the terminology of 'identification'. Similar points apply as in the previous chapter. It is suggested on the basis of the present chapter that there is no problem about taking this terminology literally in an inconsistent framework. It is a convenient way of signalling the identification relationship between two spaces, to see one as an inconsistent functional extension of the other, or for that matter an incomplete cut-down.

Let (X, O) be a topological space where O is the collection of open subsets of X, let R be an equivalence relation on X, let $X|R$ be the induced quotient set, let P be the induced projection $P : X \rightarrow X|R$, and let Q be the induced quotient topology on $X|R$ (that is, $Q =df \{S \subseteq X|R : P^{-1}(S) \in O\}$). It is not difficult to find inconsistent transparent structures which take this data into account.

To see this, let the language have as terms (i) all members of X, (ii) all subsets of X including the null set Λ, (iii) the constant term O. As before $t, t_1, t_2, \ldots$ are metalinguistic variables ranging over terms; as well, $S, S_1, S_2, \ldots$ are metalinguistic variables ranging over terms which are subsets of X. There are two binary predicates $\{=, \in\}$, and all sentences of the form $t \in S$ and $S \in O$ are stipulated to be atomic. The set of sentences is the usual closure under $\{\sim, \&, \forall\}$. Now given (X, O, R) define a model by:

(1) $I(t) = P(t) = [t]$ for all terms t in X.

(2) $I(S) = P(S) = P(R(S))$ for all terms $S \subseteq X$, where

$R(S) =df \ \{x : (\exists y)(Rxy \& y \in S)\}$.

(3) $I(t_1 = t_2) = B$ if $I(t_1) = I(t_2)$, else $I = F$.

(4) $I(t \in S) = B$ if $I(t)$ is in $I(S)$, else $I = F$.

(5) $I(S_1 = S_2) = B$ if $I(S_1) = I(S_2)$, else $I = F$.

(6) $I(S \in O) = B$ if $I(S)$ is open in $I(O)$, i.e. $I(S) \in I(O)$, else $I = F$.

This produces an inconsistent extension of the classical consistent complete theory

of the topology of (X, O), and it is straightforward to prove that I is transparent.

One could also have added into the object language terms for describing the con-

tents of $X|R$ and its topology, but the results following do not depend on the *theory*

containing descriptions of the behaviour of $X|R$, so it is better not to complicate

the issue (but it is a direction worth pursuing elsewhere). The behaviour of $X|R$

has to be taken into account metalinguistically in the statements and proofs of the

propositions following, needless to say. Also, the term-forming operations $\{\cap, \cup\}$

could be included in the object language connecting subsets of X or members of

O, and an interpretation induced with $I(S_1 \cup S_2) = I(S_1) \cup I(S_2)$ and similarly for

$\cap$, but this is not done here.

A more sophisticated model confines the inconsistency to those statements

which from the classical point of view are 'really' false. Keeping (1) and (2) as

before let

(3') $I(t_1 = t_2) = T$ if $t_1 = t_2$ in X, $I(t_1 = t_2) = B$ if $t_1 \neq t_2$ in X but $I(t_1) = I(t_2)$,

else $I = F$.

(4') $I(t \in S) = T$ if t is a member of S, $I(t \in S) = B$ if t is not a member of S

but $I(t)$ is in $I(S)$, else $I = F$.

(5') $I(S_1 = S_2) = T$ if $S_1 = S_2$ as subsets of X, $I(S_1 = S_2) = B$ if $S_1 \neq S_2$ but

$I(S_1) = I(S_2)$, else $I = F$.

(6') $I(S \in O) = T$ if S is open in O, $I(S \in O) = B$ if S is not open in O but

$I(S)$ is in $I(O)$, else $I = F$.

This $RM3$-theory fails transparency if there is even one pair t_1, t_2 in X with

Rt_1t_2 and t_1 not identical with t_2 in X : for then we have $I(t_1 = t_2) = B = I(\sim t_1 = t_2)$ and so both $t_1 = t_2$ and $\sim t_1 = t_2$ hold; so that if the theory were transparent $\sim t_1 = t_1$ would hold contradicting $I(t_1 = t_1) = T$. But the theory is functional: an atomic equation holds iff it is true when interpreted in the classical theory of the equivalence classes in $X|R$, and the latter is certainly functional. This chapter shows, hopefully, that transparency is not a particularly strong desideratum.

The class of theories considered in the rest of this chapter is obtained by taking $(1), (2), (3'), (4'), (5')$ and replacing $(6')$ by:

$(6'')$ $I(S \in O) = T$ if S is open in O, else $I(S \in O) = F$. (One could also have B instead of T without affecting the following results.)

Given the data (X, O, R) a unique structure satisfying $(1)-(6'')$ is induced, so it is named $M(X, O, R)$, or M for short. The following results show that the functionality of $M(X, O, R)$ is related to the (classical) topological properties of (X, O) and the properties of R.

Definition 10.1. The projection map $P : (X, O) \rightarrow (X|R, Q)$ is said to be *open* iff for any $S \subseteq X$ if S is open in O then $P(S)$ is open in Q. (See Kelley [16], p.94.)

Proposition 10.2. If $M(X, O, R)$ is functional then P is open.

Proof. If P is not open, then by Kelley [16] Theorem 10 p.97, there is a subset $S \subseteq X$ open in O such that $R(S)$ is not open in O. Now $I(S) = P(S) = P(R(S)) = I(R(S))$. Hence $I(R(S)) = T$ or B, that is

$$S = R(S) \text{ holds in } M \qquad (\alpha)$$

Now S is open in O, so

$$S \in O \text{ holds in } M \qquad (\beta)$$

Further, $R(S)$ is not open in O, so

$$R(S) \in O \text{ does not hold in } M. \qquad (\gamma)$$

But (α), (β) and (γ) are jointly incompatible with the functionality of M. $\square$

Definition 10.3. $O^+ =_{df} \{S \subseteq X : P(S) \text{ is open in } Q\}$; that is $O^+ = \{S \subseteq X : P^{-1}(P(S)) \text{ is open in } O\}$.

In general O^+ is not a topology, which perhaps accounts for its neglect: while X and $\wedge$ are in O^+ and O^+ is a closed under arbitrary unions, it is not always closed under finite intersections. However, the properties of O^+ are related to the topological properties of (X, O) and the functionality of $M(X, O, R)$ as we will see. One observation to make is that in general neither $O \subseteq O^+$ (see Proposition 10.5) nor $O^+ \subseteq O$ (see example following Proposition 10.11). Either of these can obtain without the other, however, as can $O = O^+$ (for example if O is the discrete or indiscrete topology), in which case O^+ is a topology.

Proposition 10.4. If $O = O^+$ then $M(X, O, R)$ is functional.

Proof. Note that if t_1, t_2 are terms in X and $t_1 = t_2$ holds in M then substitutions of t_2 for t_1 into atomic contexts are always functional: $[t_1] = [t_2]$ implies that ($[t_1]$ is in $[S]$ iff $[t_2]$ is in $[S]$). Hence if M is not functional, there must be S_1, S_2 of X such that in M, $S_1 = S_2$ and $S_1 \in O$ hold but $S_2 \in O$ does not hold. If $S_1 \in O$ holds and $O = O^+$, then S_1 is in O^+. So by definition of O^+, $P(S_1)$ is open in Q. If $S_1 = S_2$ holds in M, then $P(S_1) = P(S_2)$. Hence $P(S_2)$ is open in Q. But if $S_2 \in O$ does not hold, then S_2 is not open in O. If $O = O^+$ and S_2 is not open in O, then S_2 is not in O^+. But this is incompatible with $P(S_2)$ being open in Q. $\square$

Proposition 10.5. P is open iff $O \subseteq O^+$.

Proof. L to R : Let P be open and let S be in O. Since P is open, $P(S)$ is in Q; so S is in O^+, by definition of O^+. R to L : Suppose P is not open. Then for some $S \subseteq X$, S is in O and $P(S)$ is not in Q. By the latter, S is not in O^+; that is not $O \subseteq O^+$. $\square$

Proposition 10.6. If $M(X, O, R)$ is functional, then $O^+ \subseteq O$.

Proof. Suppose not $O^+ \subseteq O$; that is for some $S \subseteq X$, S is in O^+ and S is not in O. If S is in O^+ then $P(S)$ is in Q, so that $P^{-1}(P(S))$ is in O. That is,

$$P^{-1}(P(S)) \in O \text{ holds in } M. \tag{α}$$

Now note that $P(S) = P(P^{-1}(P(S)))$. (Reason: $[t]$ is in $P(S)$ iff t is in $P^{-1}(P(S))$ iff $[t]$ is in $P(P^{-1}(P(S)))$.) Hence, $I(S) = I(P^{-1}(P(S)))$. That is,

$$P^{-1}(P(S)) = S \text{ holds in } M. \tag{β}$$

But since S is not in O,

$$S \in O \text{ does not hold in } M. \tag{γ}$$

However (α), (β) and (γ) jointly imply that M is not functional. $\qquad\square$

Proposition 10.7. $M(X, O, R)$ is functional iff $O = O^+$.

Proof. *L to R* If M is functional, then by Proposition 10.2, P is open. So by Proposition 10.5, $O \subseteq O^+$. By Proposition 10.6, $O^+ \subseteq O$; hence $O = O^+$. *R to L* is Proposition 10.4. $\qquad\square$

Obvious examples of inconsistent functional theories, then, are those arising from the discrete topology on any X and any R; since for these $O = O^+$. (Reason: $Q = \{S \subseteq X|R : P^{-1}(S) \text{ is in } O\}$. But every $P^{-1}(S)$ is in the discrete topology O, so every S in $X|R$ is in Q. Thus if S is in O, then $P(S)$ is in Q. Hence P is open.) Again, if R is the identity relation, then for any (X, O), $(X|R, Q)$ is just an isomorphic copy of (X, O); so that $M(X, O, R)$ is functional. We see presently that there are functional theories for other topologies.

Definition 10.8. A space (X, O, R) is *R-discrete* iff for all x in X, if there is a y in X with $x \neq y$ and Rxy then the singleton $\{x\}$ is open.

R-discreteness is a kind of relativised discreteness, for example all spaces with the discrete topology are R-discrete for every R. Another property is:

Proposition 10.9. If (X, O, R) is R-discrete then P is open.

Proof. Let (X, O, R) be R-discrete and S be open in O. By Kelley [16] Theorem 10(b) p.97, it suffices to show that $R(S)$ is open in O. Let $S_1 = \{x \text{ in } S : (\exists y \text{ in } X)(Rxy \text{ and } x \neq y)\}$ and let $S_2 = S - S_1$. Then $S_1 \cup S_2 = S$ and $S_1 \cap S_2 = \Lambda$ and $R(S_2) = S_2$. Now $R(S) = R(S_1) \cup R(S_2) = R(S_1) \cup S_2$. But $S_1 \subseteq R(S_1)$. Therefore, $R(S) = (R(S_1) - S_1) \cup S_1 \cup S_2 = (R(S_1) - S_1) \cup S$. Now for all x in $R(S_1)$, there is some y in X with Rxy and $x \neq y$; so by R-discreteness $\{x\}$ is open in O. Hence for all x in $R(S_1) - S_1$ the same is true. Hence $R(S_1) - S_1$, as the union of all these $\{x\}$, is also open in O. But $R(S) = (R(S_1) - S_1) \cup S$, and S is also open in O; so $R(S)$ is open in O. $\square$

The converse of Proposition 10.9 fails, see example at the end of the chapter. The two main theorems of the chapter are the next two, which show that the presence or absence of Hausdorffness for (X, O) is relevant to the connection between R-discreteness and functionality.

Proposition 10.10. If (X, O) is a T_2 space, then $M(X, O, R)$ is functional only if (X, O) is R-discrete.

Proof. Let (X, O) be T_2, let x be in X and suppose $(\exists y)(Rxy \text{ and } x \neq y)$. It suffices to prove on the supposition of functionality that $\{x\}$ is in O. Since (X, O) is T_2, there are disjoint S_1, S_2 in O such that x is in S_1 but not S_2 and y is in S_2 but not S_1. Now since Rxy, $P(\{y\}) = P(\{x, y\})$; so $P(S_2) = P(S_2 \cup \{x\})$. Therefore, $P^{-1}(P(S_2)) = P^{-1}(P(S_2 \cup \{x\}))$. But $O^+ = \{S \subseteq X : P^{-1}(P(S)) \text{ is in } O\}$. Hence S_2 is in O^+ iff $S_2 \cup \{x\}$ is in O^+. By functionality, $O = O^+$. Hence S_2 is in O iff $S_2 \cup \{x\}$ is in O (both sides of $O = O^+$ are used here). But S_2 is in O, so $S_2 \cup \{x\}$ is in O. But S_1 is in O, so $S_1 \cap (S_2 \cup \{x\})$ is in O. S_1 and S_2 are disjoint, so $S_1 \cap (S_2 \cup \{x\}) = S_1 \cap \{x\} = \{x\}$ which is thus in O. $\square$

This proof needs all the resources of the hypothesis, in particular the Hausdorff condition that the separating open sets be disjoint, as the next Proposition shows.

Proposition 10.11. There is a space (X, O, R) such that (X, O) is T_1 but neither T_2 nor R-discrete, while $M(X, O, R)$ is functional.

Proof. Let $X = \{0, 1, 2, \ldots\}$, let $O = \{$cofinite subsets of $X\} \cup \{\wedge\}$, and let $R = \{\langle 1, 2\rangle, \langle 2, 1\rangle\} \cup \{\langle x, x\rangle : x$ in $X\}$. Now it is well known that (X, O) is T_1 but not T_2. Also, it is not R-discrete; because 1 is in X and $R12$ and $1 \neq 2$ while $\{1\}$ is not cofinite so not in O. It remains to prove that M is functional, that is $O = O^+$.

(a) $O \subseteq O^+$: This holds iff P is open, iff S is in O implies $P(S)$ is in Q, iff S is in O implies $P^{-1}(P(S))$ is in O, iff S is cofinite implies $P^{-1}(P(S))$ is cofinite. But $S \subseteq P^{-1}(P(S))$, so if S is cofinite so is $P^{-1}(P(S))$.

(b) $O^+ \subseteq O$: Let S be in O^+. Then $P^{-1}(P(S))$ is in O and so is cofinite. That is, $P^{-1}(P(S)) = X - \{x_1, \ldots, x_n\}$. Now there are two subcases:

(bi) Neither 1 nor 2 is in $P^{-1}(P(S))$. Then $P^{-1}(P(S)) = S$ and S is cofinite as required.

(bii) Both 1 and 2 are in $P^{-1}(P(S))$. (Since $R12$ there can't be one without the other.) But then $S = X - \{x_1, \ldots, x_n, 1\}$ or $S = X - \{x_1, \ldots, x_n, 2\}$ or $S = X - \{x_1, \ldots, x_n, 1, 2\}$. Whichever, S is cofinite. Thus for either (bi) or (bii), S is in O as required. $\qquad \square$

An example of a space which is T_2 with open P but neither functional nor R-discrete (thereby showing that open P is weaker than functionality and R-discreteness even given T_2) is obtained from the earlier example of the projective plane over R (see Chapter 9 end of section 3). Let $X =$ the sphere S^2, let O be the usual topology on S^2 which is T_2, let Rxy iff $x = y$ or (x, y) are an antipodal pair. Then $X|R$ constitutes the projective plane and Q is the usual topology on it. Note that S is open in O iff S' is open in O, where S' is the set of antipodes of members of S. But $R(S) = S \cup S'$; so if S is open $R(S)$ is open. That is, P is an open map. But (X, O) is not R-discrete; no singleton is an open set. Hence neither is M functional.

These results can be shown in a picture.

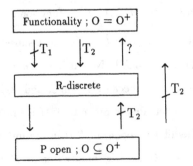

Problem: Under what conditions if any does R-discreteness imply functionality? There would also seem to be further directions to explore in this general area. Are there further conditions relating T_1 to T_2 to functionality? What of T_0 spaces? The non-T_0 space with the indiscrete topology (any X, R) is of course functional. Introducing an additional coarser topology $Q^- \subset Q$ on $X|R$ has connections with these concepts; for example, it is known that P is open only if $Q^- = Q$. The open closed duality can presumably also be exploited. The aim of this chapter, however, has been to show that there is interaction between aspects of inconsistent topological structures and their more traditional classical topological features.

CHAPTER 11: CATEGORY THEORY

(with Peter Lavers)

1. Introduction

It was claimed in Chapter 1 that the broad abstraction principles encountered in category theory look interestingly close to inconsistency. Foundational problems in category theory are well-summarised in Hatcher [15] pp.255-260, though he does not consider paraconsistency. The situation seems to be somewhat similar to that in the foundations of set theory, which is hardly surprising. A distinction can be made between large and small categories, the former being categories which, were they set-like, would contain so many sets that they would have to be proper classes (such as the category of sets). These tend to common, as Hatcher points out. The problem is that natural tendencies to abstraction lead one to want to consider several such categories as a category, and functors between them as morphisms of the category. The impulse to do so is normal category-theoretic thinking, but it appears that NBG set theory cannot make sense of it. Again, functor categories are categories whose objects are functors (between small categories perhaps) and whose morphisms are natural transformations between the functors. These appear not to be accommodated in a natural way within, say, NBG. Yet such constructions can be natural, apparently parallel to other acceptable constructions, and even required by the spirit of category theory. Needless to say several theories have been forthcoming to place the situation on a consistent footing, but these seem mostly to have various ad-hoceries, as with the more familiar case of set theory. At any rate, just as with set theory, the case for an inconsistent foundation for category theory, to allow adequately for its powerful abstraction principles, looks at least to be worth proper investigation.

However in this chapter we concentrate on a certain kind of category, namely toposes. It is well-known that set theory gives rise to Boolean algebra. It was realised by Lawvere, Grothendieck and many others that set theory could be weakened in a natural way to produce a broader class of category-theoretic structures, toposes; and that these stand to intuitionist logic, that is the logic of open sets, as sets do to classical logic. This brilliant theory proved to have many aspects. In quantification theory for example it was seen that toposes yielded a natural logic which could be described as higher-order intuitionist type theory. This is not surprising given that set theory also can naturally represent type theory, but the topos-theoretic representation is structurally deeper.

Most of this chapter is concerned with the propositional aspects of topos logic. Specifying a topological space by its closed sets is as natural as specifying it by its open sets. So it would seem odd that topos theory should be associated with open sets rather than closed sets. Yet this is what would be the case if open set logic were the natural propositional logic of toposes. At any rate, there should be a simple 'topological' transformation of the theory of toposes, which stands to closed sets and their logic, as topos theory does to open sets and intuitionism. Furthermore, the logic of closed sets is paraconsistent. This is essentially the message of Goodman's [13], though we disagree with his pessimistic conclusions, particularly concerning implication (see this chapter, section 4). There are, in fact, a number of different paraconsistent logics of closed sets, depending on different definitions of theoremhood and deducibility.

In section 2, we define paraconsistent algebras corresponding to closed set logic. In section 3 we show that a simple duality transformation of topos theory and its $\mathcal{E}$-semantics will produce such paraconsistent logics. In section 4 it is shown that there is a reasonable implication operator on these dualised toposes which produces a corresponding reasonable implication operator on the logics. This operator can also be defined in toposes, which shows that even toposes allow additional reasonable implications to the usual intuitionist implication. In section 5 we sketch quantification theory to show that intuitionism has no special claim on the quantificational aspects of the theory. The topological duality of intuitionist and paraconsistent sentential logics, as well as implication on the latter, is also considered in Mortensen and Leishman [40].

2. Closed Set Logic

Definition 11.1. *Paraconsistent algebras* are distributive lattices with a maximal element Tr, a minimal element F and a complement operation r. We suppose for convenience that these are specified by equational theories, with the order $\leq$ defined in the usual way as $a \leq b =df\ a \cap b = a$, equivalently $a \cup b = b$. In addition, paraconsistent algebras satisfy the condition that $a \cup b = Tr$ iff $ra \cup b = b$, equivalently iff $ra \cap b = ra$.

These ensure the following further properties of paraconsistent algebras.

(i) $a \cup ra = Tr$.

(ii) $rra \cup a = a$, equivalently $rra \cap a = rra$, that is $rra \leq a$.

(iii) $r(a \cap b) \leq ra \cup rb$.

(iv) $r(a \cap ra) = Tr$.

(v) $r(a \cup b) \cap (ra \cup rb) = r(a \cup b)$; that is $r(a \cup b) \leq ra \cap rb$.

(vi)   In general $a \cap ra \neq F$, and in general $a \cup rra \neq rra$, *i.e.* not $a \leq rra$; but these can be equal *e.g.* when $a = Tr$ or $a = F$.

Any closed set topology determines a paraconsistent algebra when $ra = $ the closure of the set-theoretic (Boolean) complement of the closed set a, $\cup$ and $\cap$ are set-theoretic union and intersection respectively, $Tr = $ the whole space and $F = $ the null set.

In the next two sections we deal with a propositional language, closed under conjunctions $\wedge$, disjunctions $\vee$, and paraconsistent negations r. Implication $\rightarrow$ is not included at first but is considered in section 4. In section 5 quantifiers are added.

Definition 11.2. A *paraconsistent valuation* is a function $I :$ language $\rightarrow P$ assigning atomic wffs to members of a paraconsistent algebra P and matching $\wedge$

with $\cap$, $\vee$ with $\cup$, and negations $\ulcorner$ with paraconsistent complements $\ulcorner$.

It is natural to define the consequent relation $A_1, \ldots, A_n \models B$, where the A_i and B are wffs, to mean $(\forall I)(glb\{I(A_i) : 1 \leq i \leq n\} \leq I(B))$. There are a number of options for theoremhood (semantic validity of a formula) w.r.t. a particular paraconsistent algebra:

(1) $\models A =df (\forall I)(I(A) = Tr)$;

(2) $\models A =df (\forall I)(I(A) \neq F)$;

(3) $\models A =df (\forall I)(I(A) \in D)$ where D is some proper filter, e.g. $(\forall I)(I(A) \geq t)$ for some $t \neq F$.

Corresponding to each of these there are definitions of theoremhood w.r.t. all paraconsistent algebras. If we take a paraconsistent valuation with $I(A) =$ some non-null non-universal closed set a, then $I(A \wedge \ulcorner A) =$ the boundary of $a \neq F$. If also $I(B) = F$ and $Th = \{X : I(X) \neq F\}$, then both $A, \ulcorner A \in Th$ but $B \notin Th$. (Alternatively, let $Th = \{X : I(X) \in D\}$ where D is some proper filter, *e.g.* $D = \{X : A \wedge \ulcorner A \leq X\}$.) But Th is a semitheory (respectively, theory) of any of the above logics, so they are paraconsistent. Note that $A \models B \vee \ulcorner B$ but not in general $A \wedge \ulcorner A \models B$, nor $A, \ulcorner A \models B$.

3. Propositional Logic in a Category

A knowledge of basic category-theoretic and topos-theoretic concepts is assumed here. For a clear introduction, see e.g. Goldblatt [14]. The following definition then dualises the usual definitions for toposes.

Definition 11.3. A *complement-classifier* for a category E with terminal object 1, is an object Ω together with an arrow $F : 1 \to \Omega$ satisfying the condition that for every monic arrow $f : a \rightarrowtail b$ there exists a unique arrow $\overline{X}_f$ such that

$$
\begin{array}{ccc}
a & \xrightarrow{\ f\ } & b \\
{\scriptstyle !}\downarrow & & \downarrow{\scriptstyle \overline{X}_f} \\
1 & \xrightarrow[\ F\ ]{} & \Omega
\end{array}
$$

is a pullback. $\overline{X}_f$ is the *complement-character* of f. This generalises the situation in *Set* where $F : \{0\} \to \{0,1\}$ has $F(0) = 0$, and $\overline{X}_f$ is the characteristic function of the set-complement of the image of f. That is, if $f : a \to b$ is a 1–1 function, $\overline{X}_f : b \to \{0,1\}$ is given by $\overline{X}_f = 1$ if $x \in b - f(a)$, and $\overline{X}_f(x) = 0$ if $x \in f(a)$.

An (elementary) *complement-topos* is a category with initial and terminal objects, pullbacks, pushouts, exponentiation, and a complement classifier. It is clear that, if E is a complement-topos and E' is the category obtained by renaming F as T and each $\overline{X}_f$ as X_f then E' is a topos; since initial and terminal objects, pullbacks, pushouts and exponents are prior category-theoretic notions independent of classifiers. This enables a dualisation of all topos constructions substituting F for T and $\overline{X}_f$ for X_f, as follows.

True $Tr : 1 \to \Omega$ is the complement-character of the initial object 0.

$$
\begin{array}{ccc}
0 & \xrightarrow{\ 0_1\ } & 1 \\
{\scriptstyle !}\downarrow & & \downarrow{\scriptstyle Tr \underset{df}{=} \overline{X}_{0_1}} \\
1 & \xrightarrow[\ F\ ]{} & \Omega
\end{array}
$$

This is plausible for a complement-classifier. It is the dual of the definition $\perp$ of for toposes.

Negation $\ulcorner : \Omega \to \Omega$ is the complement-character of Tr

$$
\begin{array}{ccc}
1 & \xrightarrow{\ Tr\ } & \Omega \\
{\scriptstyle !}\downarrow & & \downarrow{\scriptstyle \ulcorner \underset{df}{=} \overline{X}_{Tr}} \\
1 & \xrightarrow[\ F\ ]{} & \Omega
\end{array}
$$

This dualises $\neg$ for toposes.

Disjunction $\vee : \Omega \times \Omega \to \Omega$ is the complement-character of $\langle F, F \rangle : 1 \to \Omega \times \Omega$

$$
\begin{array}{ccc}
1 & \xrightarrow{\langle F, F \rangle} & \Omega \times \Omega \\
\downarrow{\scriptstyle !} & & \downarrow{\vee \underset{df}{=} \overline{X}_{\langle F, F \rangle}} \\
1 & \xrightarrow{\quad F \quad} & \Omega
\end{array}
$$

Compare with *Set*, where the complement of $\{(0,0)\}$ in 2×2 is $\{(1,1),(1,0),(0,1)\}$.

Conjunction $\wedge : \Omega \times \Omega \to \Omega$ is given by

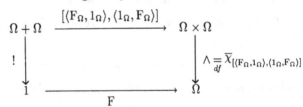

$$
\begin{array}{ccc}
\Omega + \Omega & \xrightarrow{[\langle F_\Omega, 1_\Omega \rangle, \langle 1_\Omega, F_\Omega \rangle]} & \Omega \times \Omega \\
\downarrow{\scriptstyle !} & & \downarrow{\wedge \underset{df}{=} \overline{X}_{[\langle F_\Omega, 1_\Omega \rangle, \langle 1_\Omega, F_\Omega \rangle]}} \\
1 & \xrightarrow{\quad F \quad} & \Omega
\end{array}
$$

Compare again *Set*, where the complement of conjunction is $\{(1,0),(0,1),(0,0)\}$. The above definitions of $\vee$ and $\wedge$ dualise by reversing those of $\wedge$ and $\vee$ respectively in toposes.

Let E be a complement-topos with classifier $F : 1 \to \Omega$, and let E' be the topos obtained by renaming F as $\top$ and each $\overline{X}_f$ as X_f. Let $a, b, c, \ldots$ be variables ranging over unspecified arrows of E, let S be an identity statement about E involving some of $(a, b, c, \ldots)$ as well as some subset of the constant arrows $(F, Tr, \ulcorner, \vee, \wedge)$, and let S' be the statement obtained by substituting $(\top, \bot, \neg, \wedge, \vee)$ respectively for the latter. Then

Proposition 11.4. (Duality Theorem) S is true in E if S' is true in E'.

Proof. It is clear that the diagrams for $(F, Tr, \ulcorner, \vee, \wedge)$ are diagrams for $(\top, \bot, \neg, \wedge, \vee)$ where these are renamed, and that compositions, pullbacks, pushouts, initial and terminal objects and exponents are prior category-theoretic notions unaffected by the renaming. So any construction establishing identity in E is under the renaming a construction establishing identity in E', and vice versa. $\square$

Definition 11.5. As for the usual $\mathcal{E}$-semantics for toposes, the *truth values* of a complement-topos E are the monics: $1 \to \Omega$, also called *elements* of Ω. A *paraconsistent valuation* on E is a function $I : \text{language} \to \Omega$, in which the atomic wffs are assigned to truth values and the connectives $\{\neg, \vee, \wedge\}$ agree with their complement-topos counterparts.

Proposition 11.6. The truth values of E form a paraconsistent algebra, when Tr is interpreted as the maximal element and F is interpreted as the minimal element, and $\wedge, \vee$ are interpreted as $\cap$ and $\cup$ respectively.

Proof. This is a matter of verifying that the conditions for a paraconsistent algebra in section two are satisfied, and these are ensured by the duality theorem. First one needs $a \cup b = a$ iff $a \cap b = b$; but the dual is $a \cap b = a$ iff $a \cup b = b$ which holds for open sets and for the elements of Ω in topos theory. For latticehood one needs first the partial order properties (1)–(3):

(1) reflexivity, *i.e.* $a \leq a$, *i.e.* $a \cup a = a \cap a = a$

(2) antisymmetry, *i.e.* $(a \cup b = b$ and $a \cup b = a)$ only if $a = b$

(3) transitivity, *i.e.* $(a \cup b = b$ and $b \cup c = c)$ only if $a \cup c = c$. Also:

(4) $a \cap b \leq a, b$, *i.e.* $(a \cap b) \cap a = (a \cap b) \cap b = a \cap b$

(5) $c \leq a \cap b$ iff $c \leq a$ and $c \leq b$, *i.e.* $(a \cap b) \cap c = c$ iff $(a \cap c = c$ and $b \cap c = c)$

(6) $a, b \leq a \cup b$, *i.e.* $a \cup (a \cup b) = b \cup (a \cup b) = a \cup b$

(7) $a \cup b \leq c$ iff $(a \leq c$ and $b \leq c)$, *i.e.* $(a \cup b) \cup c = c$ iff $(a \cup c = c$ and $b \cup c = c)$.

Dualised, these are respectively

(1d) $a \cap a = a \cup a = a$

(2d) $(a \cup b = b$ and $a \cap b = a)$ only if $a = b$

(3d) $(a \cap b = b$ and $b \cap c = c)$ only if $a \cap c = c$

(4d) $(a \cup b) \cup a = (a \cup b) \cup b = a \cup b$

(5d) $(a \cup b) \cup c = c$ iff $(a \cup c = c$ and $b \cup c = c)$

(6d) $a \cap (a \cap b) = b \cap (a \cap b) = a \cap b$

(7d) $(a \cap b) \cap c = c$ iff $((a \cap c) = c$ and $(b \cap c) = c)$.

These are all facts about truth values in a topos E', indeed facts about lattices of open sets. For distributivity, one needs (8) $a \cup (b \cap c) = (a \cup b) \cap (a \cup c)$ and $a \cap (b \cup c) = (a \cap b) \cup (a \cap c)$; which dualise to (8d) $a \cap (b \cup c) = (a \cap b) \cup (a \cap c)$ and $a \cup (b \cap c) = (a \cup b) \cap (a \cup c)$ which both hold of the elements of toposes. The maximality of Tr and minimality of F are (9) $a \cup Tr = Tr$ and $a \cap F = F$

respectively; which dualise to (9d) $a \cap \bot = \bot$ and $a \cup \top = \top$. These assert the minimality of $\bot$ and the maximality of $\top$ and hold of E' as desired. Finally one needs (10) $a \cup b = Tr$ iff $\ulcorner a \cup b = b$. Dualised this is (10d) $a \cap b = \bot$ iff $\neg a \cap b = b$, i.e. $a \cap b = \bot$ iff $b \leq \neg a$; but this is the condition for pseudo-complements in the topos E', and holds of open sets when $\neg a$ is the interior of the Boolean complement of a and $\bot$ is the null set (Goldblatt [14], p.179).   □

One can go to verify that the consequent further properties (i)–(vi) of paraconsistent algebras specified in section 2 also hold of complement-toposes.

(i) $a \cup \ulcorner a = Tr$ dualises to $a \cap \neg a = \bot$ which holds of the truth values of a topos as well as of open sets.

(ii) $\ulcorner\ulcorner a \cup a = a$, i.e. $\ulcorner\ulcorner a \leq a$, dualises to $\neg\neg a \cap a = a$, i.e. $a \leq \neg\neg a$, which holds of open sets.

(iii) $\ulcorner(a \cap b) = \ulcorner a \cup \ulcorner b$ dualises to $\neg(a \cup b) = \neg a \cap \neg b$ which holds of open sets.

(iv) $\ulcorner(a \cap \ulcorner a) = Tr$ dualises to $\neg(a \cup \neg a) = \bot$; the Boolean complement of $a \cup \neg a$ is a boundary, so its interior is the null set.

(v) $\ulcorner(a \cup b) \cap (\ulcorner a \cap \ulcorner b) = \ulcorner(a \cup b)$, i.e. $\ulcorner(a \cup b) \leq \ulcorner a \cap \ulcorner b$, dualises to $\neg(a \cap b) \cup (\neg a \cup \neg b) = \neg(a \cap b)$, i.e. $\neg a \cup \neg b \leq \neg(a \cap b)$, which holds of open sets.

(vi) $a \cap \ulcorner a \neq F$, and $a \cup \ulcorner\ulcorner a \neq \ulcorner\ulcorner a$ i.e. not $a \leq \ulcorner\ulcorner a$, dualise to $a \cup \neg a \neq \top$ and $a \cup \neg\neg a \neq \neg\neg a$ i.e. not $\neg\neg a \leq a$, and these are in general inequalities for open sets.

But the inequalities become equalities when $a = Tr$ or $a = F$; since $Tr \cap \ulcorner Tr = F$, $Tr \cup \ulcorner\ulcorner Tr = \ulcorner\ulcorner Tr$, $F \cap \ulcorner F = F$ and $F \cup \ulcorner\ulcorner F = \ulcorner\ulcorner F$ dualise respectively to $\bot \cup \neg\bot = \top$, $\bot \cap \neg\neg\bot = \neg\neg\bot$, $\top \cup \neg\top = \top$ and $\top \cap \neg\neg\top = \neg\neg\top$, all of which hold in toposes.

Thus we now have from the previous two propositions and according to the definitions of theoremhood and deducibility of section 2.

Proposition 11.7. The set of all paraconsistent valuations on a complement-topos determines a paraconsistent logic.   □·

4. Implication

It is time to come clean about implication, on which the authors of this chapter hold slightly different views. The usual intuitionist story defines $\Rightarrow: \Omega \times \Omega \to \Omega$ as χ_e, where e is the equaliser $e: \textcircled{\scriptsize$\leq$} \rightarrowtail \Omega \times \Omega \overset{\cap}{\underset{pr_1}{\rightrightarrows}} \Omega$; and then for any arrows f, g which are truth values: $1 \to \Omega$, defines $f \Rightarrow g$ to be $\Rightarrow \circ \langle f, g \rangle$, where $\langle f, g \rangle$ is the usual product map: $1 \to \Omega \times \Omega$.

We note here that in toposes there also exists e' as the equaliser $e': \textcircled{\scriptsize$\geq$} \rightarrowtail \Omega \times \Omega \overset{\cup}{\underset{pr_1}{\rightrightarrows}} \Omega$, as well as versions substituting pr_2 for pr_1. In the corresponding complement-topos, the same constructions obviously exist, with the property that $\langle f, g \rangle$ factors through e in the topos iff $\langle f, g \rangle$ factors through e' in the complement-topos, and similarly with e and e' reversed.

PL thinks that the right dualisation of χ_e in the topos is $\overline{\chi}_{e'}$ in the complement-topos. $\overline{\chi}_{e'} \circ \langle f, g \rangle$ corresponds to Curry's $g - f$; but this produces $f \to f = F$, for every truth value f. It isn't obvious how to avoid this consequence if one has $\overline{\chi}$ without χ. PL favors having both; which strictly takes one outside toposes, but has the advantage of allowing a truth range and a falsity range for every concept. CM thinks that there is at least one simple and reasonable implication on any lattice, namely $f \Rightarrow g = Tr$ if $f \leq g$, else $f \Rightarrow g = F$. This can be produced in complement toposes as follows.

Proposition 11.8. Let f and g be truth values $f, g: 1 \to \Omega$. Then the product map $\langle f, g \rangle$ is exactly one of two types.

Type 1. The domain d of the pullback arrow $\langle i, j \rangle: d \to \textcircled{\scriptsize$\leq$}$ is not isomorphic with 1.

$$
\begin{array}{ccc}
d & \overset{\langle i, j \rangle}{\rightarrowtail} & \textcircled{\scriptsize$\leq$} \\
{\scriptstyle !}\downarrow & \text{PB} & \downarrow {\scriptstyle e} \\
1 & \underset{\langle f, g \rangle}{\rightarrowtail} & \Omega \times \Omega
\end{array}
$$

Type 2. $\langle f, g \rangle$ factors through $e : \text{ⓢ} \rightarrowtail \Omega \times \Omega$

$$
\begin{array}{ccc}
& & \text{ⓢ} \\
\langle \text{f}, \text{g} \rangle_F \nearrow & & \downarrow e \\
1 \rightarrowtail & \Omega \times \Omega \\
& \langle \text{f}, \text{g} \rangle
\end{array}
$$

where $\langle f, g \rangle_F$ is the factorisation of $\langle f, g \rangle$ through e.

Proof. Consists of proving that $\langle f, g \rangle$ is of type 2 iff $d \cong 1$ in diagram 1.

L to R : Draw the diagram:

$$
\begin{array}{ccc}
& \langle \text{i}, \text{j} \rangle & \\
d \rightarrowtail & & \text{ⓢ} \\
\Big\downarrow ! & \langle \text{f}, \text{g} \rangle_F \nearrow & \Big\downarrow e \\
1 \rightarrowtail & \Omega \times \Omega \\
& \langle \text{f}, \text{g} \rangle
\end{array}
$$

It is desired to prove that ! is iso. Now $\langle f, g \rangle \circ ! = e \circ \langle i, j \rangle$ by the pullback. But $\langle f, g \rangle = e \circ \langle f, g \rangle_F$ by the factorisation. So $e \circ \langle f, g \rangle_F \circ ! = e \circ \langle i, j \rangle$. But e is monic, so left-cancellable; hence $\langle f, g \rangle \circ ! = \langle i, j \rangle$. But $\langle i, j \rangle$ is monic since $\langle f, g \rangle$ is, by the pullback. So $\langle f, g \rangle \circ !$ is monic. But then ! is monic (Goldblatt p.39 Ex(2)). Now ! is certainly epic, so ! is both epic and monic. But in any complement-topos, as in any topos, an epic monic is iso (proof of this by running through Goldblatt pp.109-110 substituting F for T and $\overline{X}_f$ for X_f, and then for the Corollary noting that Theorem 2 p.57 is independent of the dualisation). Thus $d \cong 1$.

R to L : If ! is iso then it has an inverse $!^{-1}$. Then $\langle f, g \rangle_F$ can be defined as $\langle i, j \rangle \circ !^{-1}$. This is because the arrow $\langle i, j \rangle \circ !^{-1} : 1 \rightarrow \text{ⓢ}$ is unique in making the diagram commute: if $k : 1 \rightarrow \text{ⓢ}$ is any other arrow making the diagram commute, then we have that $\langle f, g \rangle = e \circ \langle i, j \rangle \circ !^{-1} = e \circ k$; but since e is monic and left-cancellable, $k = \langle i, j \rangle \circ !^{-1}$ as required for the definition. That is, $\langle f, g \rangle$ factors through ⓢ, so $\langle f, g \rangle$ is of type 2. $\qquad \square$

Now let f_1, g_1, f_2, g_2 be truth values such that $\langle f_1, g_1 \rangle$ is type 1 and $\langle f_2, g_2 \rangle$ is type 2. (Note that there is at least one of each type: $\langle Tr, F \rangle$ is type 1 and $\langle F, Tr \rangle$

is type 2.) Draw the composite diagram:

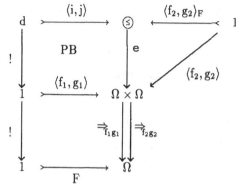

where $d \not\cong 1$; and where $\Rightarrow_{f_1 g_1}$ is defined to be $\overline{X}_{\langle f_1, g_1 \rangle}$ and $\Rightarrow_{f_2 g_2}$ is defined to be $Tr_{\Omega \times \Omega}$, i.e. $Tr \circ 1_{\Omega \times \Omega}$, where $1_{\Omega \times \Omega} : \Omega \times \Omega \to 1$. (Note that $\Rightarrow_{f_1 g_1}$ could be defined as $F_{\Omega \times \Omega}$ with the same result.)

Definition 11.9. Let f, g be any truth values. Then

$$f \Rightarrow g =df \ \Rightarrow_{fg} \circ \langle f, g \rangle.$$

Note that $f \Rightarrow g : 1 \to \Omega$ is a truth value, and the $\Rightarrow_{fg} : \Omega \times \Omega \to \Omega$.

Proposition 11.10. $f \Rightarrow g = Tr$ if $f \leq g$; else $f \Rightarrow g = F$.

Proof. If $f \leq g$, then $\langle f, g \rangle$ factors through e; hence $\Rightarrow_{fg} = Tr_{\Omega \times \Omega}$, and $f \Rightarrow g = Tr_{\Omega \times \Omega} \circ \langle f, g \rangle$. The latter can be shown to equal Tr as follows:

$$Tr_{\Omega \times \Omega} \circ \langle f, g \rangle = (Tr \circ 1_{\Omega \times \Omega}) \circ \langle f, g \rangle = Tr \circ (1_{\Omega \times \Omega} \circ \langle f, g \rangle).$$

Now $1_{\Omega \times \Omega} \circ \langle f, g \rangle : 1 \to 1$; but there is a unique arrow $1 \to 1$, viz 1_1. So

$$f \Rightarrow g = Tr \circ 1_1 = Tr.$$

Else $\langle f, g \rangle$ is of the type 1 and $\Rightarrow_{fg} = \overline{X}_{\langle f, g \rangle}$; so that $f \Rightarrow g = \overline{X}_{\langle f, g \rangle} \circ \langle f, g \rangle = F$. $\square$

There is obviously an identical account of implication for toposes, so it is not even true that toposes support a single account of implication, even setting aside the existence of complement-toposes. The present account is reasonable from at least two points of view; it is a reasonable implication, and its category-theoretic account is not too complex. Also the latter is reasonably general in that it can obviously be easily modified to produce other implications.

We stress that the importance of implication can be over estimated: implication is an object-language expression of the metalinguistic $\models$, which is always available for $\{\vee, \wedge, \ulcorner\}$ logic as in section 2. These constructions also bear out the fact that some natural implications such as the intuitionist X_e are sensitive to the classifier; while others such as that of the present section have deeper, topological invariance. It seems, further, that it is the topological aspects which support intuitionist and paraconsistent theories as alternatives to classical theories.

5. Quantification Theory

Our main concern has not been quantification theory; if only because of its extent, variety and complexity, but also because it is the topological duality of intuitionism and paraconsistency which mostly interests us; but something should be said. It seems to us that there are no further substantial changes to be made to obtain a reasonable quantifier logic. Elementary quantification logic needs a re-interpretation in the light of the duality. It is seen that this is achieved by replacement of $(\forall_a, \exists_a)$ by $(\exists_a, \forall_a)$ throughout, as might be expected given their roles as generalised conjunction and disjunction. Higher-order logic seems to be independent of the dualisation, since it derives from Cartesian closedness of the category, which is prior to the dualisation. Again this is not unexpected given that set theory supports classical type theory. These points are expanded on briefly below.

This section follows the admirable exposition in Goldblatt [14] pp.238-248. One has for toposes (i) $\forall_a b = T$ if $a = b$, else $\forall_a b \neq T$, and (ii) $\exists_a b = T$ if $b \neq \Lambda$, else $\exists_a b \neq T$; where a is any object in the category and functions as the domain of quantification or universal set for the quantifiers and Λ is the null set. The elements of a are the morphisms $x : 1 \rightarrow a$ which provide the interpretation of terms; m-place predicates (atomic or complex) are morphisms from a^m to Ω, with a distinguished set of elements namely those mapped to T. Closed sentences take values in Ω which is a complete Heyting algebra, and then are combined by the intuitionist sentential operators.

A diagram can be constructed.

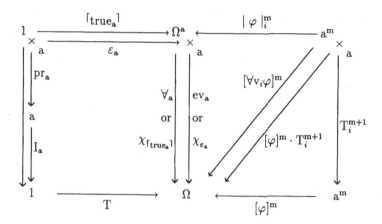

φ^m is an arbitrary m-adic predicate, $[\varphi]^m$ is its semantic interpretation as a map from $a^m \to \Omega$, and $\ulcorner true_a \urcorner$ and $|\varphi|_i^m$ are exponential adjoints of $true_a \circ pr_a$ and $[\varphi]^m \circ T_i^{m+1}$ respectively. $\forall_a$ is defined to be the character of $\ulcorner true_a \urcorner$. Moving to complement-toposes, it is reasonable to take designated values to be all elements other than F, as mentioned in section 2. Replacing T by the F of complement-toposes, renaming $true_a$ as $false_a$ and $\ulcorner true_a \urcorner$ as $\ulcorner false_a \urcorner$, and renaming $\forall_a$ as $\exists_a$, one has by the same construction as in toposes, that $\exists_a b = F$ if $a = b$, else $\exists_a b \neq F$. Now this is reasonable if b is the set of elements of a for which φ is 'absolutely false' or perhaps 'consistently false', that is F. That is to say, $\exists_a v_i \varphi^m$ is F if the *counter-extension* b of φ is identical with a. Thus $\exists_a$ is the *complement-character* of $\ulcorner false_a \urcorner$. Similarly, in toposes $\exists_a$ is the character of the image of the composite $p_a \circ \varepsilon_a$, where p_a is the first projection arrow : $\Omega^a \times a \to \Omega^a$. Renaming this as $\forall_a$ in complement-toposes, one obtains $\forall_a b = F$ if $b \neq \Lambda$, else $\forall_a b \neq F$. Again this is reasonable if b is the set of elements of a of which φ is absolutely false. That is, $\forall_a v \varphi$ is F when the set of elements of a of which φ is consistently false is non-null, else $\forall_a v \varphi$ is non-F.

There are very many aspects of quantification theory of toposes which we do not deal with, but identity and higher order types deserve mention. It is customary to treat identity as a logical symbol; so that it is built into the assignment clause for identity that $\ulcorner t = t$ does not hold. In light of the earlier chapters, this is not natural for complement-toposes, and so one should discard assignment clauses for identity such as that on Goldblatt p.246. Higher order logic and type theory is obtained in topos theory because of the fact that toposes, being Cartesian closed categories, contain all powerobjects, which serve as semantic values of syntactic types. However, powerobjects are prior to the classifier, so it is to be expected that these constructions go over to complement-toposes unchanged. This is perhaps not so surprising given that a parallel construction can be done for classical logic and set theory with powersets.

6. Conclusion

It hardly bears saying that a complement-topos really is a topos, it is just a matter of how one understands the notion of a subobject classifier. This can be masked by the usual terminology of $T : 1 \to \Omega$, rather than F as we have. There are various natural paraconsistent propositional logics arising from the $\mathcal{E}$-semantics of complement-toposes, and thus from toposes. The bias toward intuitionism is at least not justified by these structural aspects of a topos, since it depends on how they are interpreted.

We conclude this section by raising the question of whether there is a construction internal to topos theory (rather than dualising 'outside' the topos as we have) which also yields paraconsistent logic. This seems not unreasonable. It also seems reasonable to think that there are other avenues of the dualisation; for example closed set sheaves, which are accordingly explored in the next chapter.

CHAPTER 12: CLOSED SET SHEAVES AND THEIR CATEGORIES

(by William James)

1. Introduction

With this chapter we examine categorial sheaves over the closed sets of a topological space. Aside from the historical interest that sheaves seem to have been defined over closed sets first before the more usual definition over open sets, there are a number of other reasons for developing the theory of sheaves over closed sets. First of all having a base topology of closed sets gives us a working concept of boundary that does not exist for the open set sheaf notion. One area in which this may work for us is the mathematics of physics. Lawvere in the introduction to *Categories in Continuum Physics* [21] mentions the speculation that there is a role for a closed set sheaf in thermodynamics as a functor from a category of parts of a body to a category of "abstract thermodynamical state-and-process systems" (p.9). Another reason for closed set sheaves is their effect on categorial logic. A closed set topology ordered by set inclusion is a paraconsistent algebra. Via the sheaves we can introduce this paraconsistency to toposes.

The first two sections contain a brief description of some of the existing theory of categorial sheaves. We note that categories of sheaves as standardly understood are toposes. It will be our claim that categories of SET-valued sheaves over the closed sets of a topological space are toposes in just the same way. Since the existence of a subobject classifier and the resulting subobject classifying maps is a defining feature of a topos, we will be obliged to show, contrary to some standard presentations, that there is a construction for the classifying arrows χ of sheaf monics that does not rely on $\bigcup$-completeness of the base space topology. We establish the necessary construction as a corollary to a theorem at the end of section two. This clears the way for section three where we briefly justify the notion of a sheaf over closed sets and the claim that the category of all sheaves over the closed sets of topological space is a topos.

In essence what we describe is a category, indeed a topos, of sheaves over a closed set topology understood as a poset category. It should be noted that our discussion ultimately is given in terms of the theory of j-sheaves and as a result the categories we describe are not necessarily equivalent to categories of continuous local homeomorphisms or sheaf spaces. We therefore emphasise a particular type of j-sheaf category: one that includes a set theoretic covering system for the base topology. It will then be appropriate to describe our particular construction as a category of sheaves.

For a category $\mathcal{C}$ any contravariant functor $\mathcal{C}^{op} \to \text{SET}$ is called a SET-valued *presheaf*. The SET-valued *sheaves* are a special subset of the presheaves. Throughout this discussion we assume that $\mathcal{C}$ is a small category.

2. Pretopologies and Topologies for Categories

This section follows similar discussions in Johnstone [15a] and in Goldblatt [14].

Definition 12.1. A *pretopology* on a category $\mathcal{C}$ with pullbacks is a system P where for each $\mathcal{C}$-object U there is a set $P(U)$. Each $P(U)$ contains families of $\mathcal{C}$-morphisms $\{U_i \xrightarrow{\alpha_i} U : i \in I\}$. The following conditions are satisfied.

(i)   for each $U \in \mathcal{C}$, $\{id_U\} \in P(U)$;

(ii)   if $V \to U$ in $\mathcal{C}$ and $\{U_i \xrightarrow{\alpha_i} U : i \in I\} \in P(U)$, then $\{V \times_U U_i \xrightarrow{\pi_1} V : i \in I\}$ in $P(V)$. Note π_1 is the pullback in $\mathcal{C}$ of α_i along $V \to U$;

(iii) if $\{U_i \xrightarrow{\alpha_i} U : i \in I\} \in P(U)$, and for each $i \in I$, $\{V_{ik} \xrightarrow{\beta_{ik}^i} U_i : k \in K_i\} \in P(U_i)$, then $\{V_{ik} \xrightarrow{\beta_{ik}^i} U_i \xrightarrow{\alpha_i} U : i \in I, k \in K_i\} \in P(U)$. Note that V_{ik} is an example of a double indexed object rather than the intersection of V_i and V_k.

The notion of a pretopology is a categorial generalisation of a system of (set theoretic) covers on a topology $\mathcal{T}$ where a *cover* for $U \in \mathcal{T}$ is a set $\{U_i : U_i \in \mathcal{T}, i \in I\}$ such that $\bigcup \{U_i : i \in I\} = U$. The generalisation is achieved by noting that the topology ordered by inclusion is a (poset) category and that any cover corresponds to a collection of inclusion arrows $U_i \to U$. Given this, any family of arrows contained in $P(U)$ of a pretopology is also called a cover.

Definition 12.2. In analogy with sheaves over a topological space we have the notion of sheaves over categories with pretopologies. We shall say that any contravariant functor $F : \mathcal{C}^{op} \to \mathrm{SET}$ is a *sheaf* just in case for each $U \in \mathcal{C}$ and for each $\{U_i \xrightarrow{\alpha_i} U : i \in I\} \in P(U)$, we have an equaliser

$$F(U) \longrightarrow \prod_{i \in I} F(U_i) \underset{d_1}{\overset{d_0}{\rightrightarrows}} \prod_{i,j} F(U_i \times_U U_j)$$

where d_0 and d_1 are product arrows determined respectively by the images under F of the first and second projection maps $U_i \times_U U_j \to U_i$ and $U_i \times_U U_j \to U_j$, all $i, j \in I$.

Pretopologies do not in general uniquely determine a category of sheaves. To do that we need the notion of a (categorial) topology.

Definition 12.3. For an object U in a category $\mathcal{C}$ a U-*sieve* is a family R of $\mathcal{C}$-morphisms with codomain U such that if $V \xrightarrow{\alpha} U \in R$ and $W \xrightarrow{\beta} V$ is any $\mathcal{C}$-morphism, then the composite $W \xrightarrow{\beta} V \xrightarrow{\alpha} U \in R$. (Some writers make no terminological distinction between sieves and their categorial duals. Others do, and name the duals *cribles*. Still others, notably Goldblatt [14], use the opposite naming convention, their crible being our sieve). A *topology* on $\mathcal{C}$ is a system J of sets $J(U)$ for each $U \in \mathcal{C}$ where each $J(U)$ is a set of U-sieves called *covering* sieves. We have the following conditions:

A system J is a topology for C if:

(i)   for any $U \in C$, the *maximal* sieve $\{\alpha : \text{cod}(\alpha) = U\} \in J(U)$;

(ii)   if $R \in J(U)$ and $V \xrightarrow{f} U$ is a morphism of C, then $f^*(R) = \{W \xrightarrow{\alpha} V : f \cdot \alpha \in R\}$ is in $J(V)$;

(iii)   if $R \in J(U)$ and S is a sieve on U where for each $(V \xrightarrow{f} U) \in R$ we have $f^*(S)$ in $J(V)$, then $S \in J(U)$.

Note that a collection of morphisms with codomain U can be a U-sieve without being a covering sieve on U.

A small category C together with a topology J is called a *site*. We now define a *sheaf on a site* (C, J) to be any contravariant functor $F : C^{op} \to$ SET satisfying the equaliser condition expressed in terms of covering sieves for U rather than covers. A category of sheaves on a site is called a *Grothendieck topos* and denoted $sh(C, J)$.

Proposition 12.4. (Johnstone, [15a]) Any Grothendieck topos is an elementary topos.   □

Proposition 12.5. Given a pretopology P we can define a topology J that will give rise to the same sheaves on C. We say that for any $U \in C$, we have $R \in J(U)$ iff R contains a pretopology cover $\{\alpha_i : i \in I\} \in P(U)$.   □

Definition 12.6. The category of all presheaves on C is denoted $\text{SET}^{C^{op}}$ and when C is a small category, $\text{SET}^{C^{op}}$ is a topos (Goldblatt, [14], pp.204-210). The *classifier object* in $\text{SET}^{C^{op}}$ is a presheaf $\Omega : C^{op} \to$ SET where for $U \in C$,

$$\Omega(U) = \{\text{all sieves on } U\},$$

and for $V \xrightarrow{f} U$ in C, $\Omega(f) : \Omega(U) \to \Omega(V)$ (also denoted Ω_V^U) is given by

$$\Omega(U) \ni S \mapsto \{W \xrightarrow{\alpha} V : f \cdot \alpha \in S\} \in \Omega(V).$$

When all arrows in C are inclusions this becomes $S \mapsto \{W : W \subseteq V \text{ and } W \in S\}$.

A topology J exists as a presheaf $J : \mathcal{C}^{op} \to \text{SET}$ where we have $\mathcal{C} \ni U \mapsto J(U)$ and for any $V \xrightarrow{f} U$ in $\mathcal{C}$ the map $J(f) : J(U) \to J(V)$ is given by $R \mapsto f^*(R)$. Clearly, J is a subobject of Ω; that is, an inclusion $J \to \Omega$ exists. The classifying map associated with this inclusion is denoted by j, and since it has proven possible to describe the same category of sheaves on $\mathcal{C}$ in terms of either J or j, that map $j : \Omega \to \Omega$ is also called a topology. Note that J is a topology on $\mathcal{C}$ and j is a topology in $\text{SET}^{\mathcal{C}^{op}}$.

The notion of a topology as an endomorphism of the classifier object has been extended to all elementary toposes.

Definition 12.7. Any map $j : \Omega \to \Omega$ in a topos $\mathcal{E}$ is a *topology in $\mathcal{E}$* if the following conditions are met:

(i) $j \cdot \text{true} = \text{true}$;

(ii) $j \cdot j = j$;

(iii) $\cap \cdot (j \times j) = j \cdot \cap$.

Sheaves are then distinguished objects of $\mathcal{E}$ identified with respect to j. Such objects are called *j-sheaves*. A monic $X' \xrightarrow{\alpha} X \in \mathcal{E}$ is called *j-dense* if its classifying map χ_α factors through $J \rightarrowtail \Omega$.

Proposition 12.8. For any topos $\mathcal{E}$, an object F is a j-sheaf if and only if for any $\mathcal{E}$-arrow $\beta : X' \to F$ and any j-dense monic $\alpha : X' \rightarrowtail X$ there is exactly one $\beta' : X \to F$ such that the following diagram commutes:

The category of sheaves identified in this manner is a full sub-category of $\mathcal{E}$ and will be denoted $sh_j(\mathcal{E})$.

Proposition 12.9. (Johnstone, [15a]) If $\mathcal{E}$ is a topos with topology $j : \Omega \to \Omega$, then $sh_j(\mathcal{E})$ is also a topos.   □

3. Subobject Classifiers

We have seen described in Definition 12.6 the standard construction of the classifier object for SET-valued presheaf categories. We complete the description by giving the cosntruction for the classifier arrow *true* and subobject classifying arrows. We follow this with a dependent construction for the subobject classifier of a j-sheaf category. We include a theorem about that construction.

Definition 12.10. The map $true: 1 \to \Omega$ in $\mathrm{SET}^{\mathcal{C}^{op}}$ is a natural transformation given by components $true_a$ for all $a \in \mathcal{C}$. The functor 1 is given by $\mathcal{C} \ni U \mapsto \{\emptyset\}$ with the obvious restriction maps. Clearly this is a terminal object for $\mathrm{SET}^{\mathcal{C}^{op}}$. The components of *true* are $true_a : \{\emptyset\} \to \Omega(a)$ where $true_a(\emptyset) = $ maximal a-sieve. Equally, 1 is a j-sheaf for any j (trivially true by Proposition 12.8) and will be terminal for the sheaf category $sh_j(\mathrm{SET}^{\mathcal{C}^{op}})$.

For any $\mathrm{SET}^{\mathcal{C}^{op}}$-monic $\tau : F \rightarrowtail G$ the classifying arrow χ_τ is a natural transformation $G \to \Omega$ given by components $(\chi_\tau)_a : G(a) \to \Omega(a)$ such that for any $x \in G(a)$, we have

$$(\chi_\tau)_a(x) = \{b \to a : G_b^a(x) \in \tau_b(F(b))\}$$

where $b \to a$ is a $\mathcal{C}$-morphism, G_b^a is the restriction map $G(a) \to G(b)$, and τ_b is the b-component $F(b) \to G(b)$ of the natural transformation τ. It is straightforward to confirm that χ_τ is a natural transformation and is the unique map that makes the following a pullback in $\mathrm{SET}^{\mathcal{C}^{op}}$.

$$
\begin{array}{ccc}
F & \stackrel{\tau}{\rightarrowtail} & G \\
\downarrow & & \downarrow{\scriptstyle \chi_\tau} \\
1 & \underset{true}{\longrightarrow} & \Omega
\end{array}
$$

Proposition 12.11 (as taken from [14], p.371). The category $sh_j(\text{SET}^{\mathcal{C}^{op}})$ has a subobject classifier and it can be described by the following $\text{SET}^{\mathcal{C}^{op}}$-diagram where e is an equaliser and $true_j$ is the unique arrow making the diagram commute:

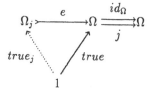

So, given that we are dealing with SET-valued functors, for all $a \in \mathcal{C}$, we have

$$\Omega_j(a) = \{S : S \in \Omega(a) \quad \text{and} \quad (id_\Omega)_a(S) = j_a(S)\}. \qquad \square$$

We intend now to show that the classifying maps, χ^j_τ, for monics, $\tau : F \rightarrowtail G$ in $sh_j(\text{SET}^{\mathcal{C}^{op}})$ are similarly related to classifying maps for monics in $\text{SET}^{\mathcal{C}^{op}}$. First we need

Proposition 12.12. (Johnstone, [15a]) For any topos $\mathcal{E}$ with topology j, the category $sh_j(\mathcal{E})$ has finite limits and the inclusion functor $sh_j(\mathcal{E}) \to \mathcal{E}$ preserves them. $\qquad \square$

In essence, the limit in $\mathcal{E}$ of a finite diagram of j-sheaves is a j-sheaf. And in particular a pullback in $sh_j(\text{SET}^{\mathcal{C}^{op}})$ is a pullback in $\text{SET}^{\mathcal{C}^{op}}$ and a pullback in $\text{SET}^{\mathcal{C}^{op}}$ of j-sheaves is a pullback in $sh_j(\text{SET}^{\mathcal{C}^{op}})$. As a corollary, any $sh_j(\mathcal{E})$-monic is monic in $\mathcal{E}$ and any morphism between j-sheaves that is monic in $\mathcal{E}$ is monic in $sh_j(\mathcal{E})$. This holds since any map $A \xrightarrow{u} B$ is monic if and only if the following is a pullback.

$$
\begin{array}{ccc}
A & \xrightarrow{id_u} & A \\
{\scriptstyle id_u}\downarrow & & \downarrow{\scriptstyle u} \\
A & \xrightarrow{u} & B
\end{array}
$$

Proposition 12.13. When $\mathcal{E}$ is a topos with topology j and $\tau : F \rightarrowtail G$ is a $sh_j(\mathcal{E})$-monic, we have $\chi_\tau = e \cdot \chi_\tau^j$.

Proof. Consider the diagram

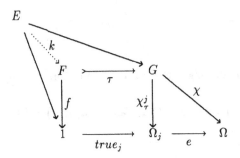

Let τ be a monic and the inner square be a pullback in $sh_j(\mathcal{E})$. Let $\chi = e \cdot \chi_\tau^j$. We will refer to parts of the above diagram by (clockwise) vertices. So, $\{F, G, \Omega_j, 1\}$ is the inner square, that is, the pullback diagram for $true_j$ and χ_τ^j.

Map e is an equaliser and therefore monic, so

$$\chi_\tau^j \cdot \tau = \text{true}_j \cdot f \text{ iff } e \cdot \chi_\tau^j \cdot \tau = \chi \cdot \tau = e \cdot \text{true}_j \cdot f$$

(The conditional, if LHS then RHS, is trivial). In other words $\{F, G, \Omega_j, 1\}$ commutes iff $\{F, G, \Omega, 1\}$ commutes. Likewise, we prove that $\{E, G, \Omega_j, 1\}$ commutes iff $\{E, G, \Omega, 1\}$ commutes and indeed that $\{F, G, \Omega_j, 1\}$ is a pullback iff $\{F, G, \Omega, 1\}$ is a pullback. But $\{E, G, \Omega_j, 1\}$ is a pullback; and since $e \cdot true_j = true$, there is, by definition, exactly one χ that makes $\{E, G, \Omega, 1\}$ a pullback in $\mathcal{E}$, namely χ_τ. So, $e \cdot \chi_\tau^j = \chi_\tau$. $\square$

Corollary. For SET-valued j-sheaves over a category $\mathcal{C}$, any $a \in \mathcal{C}$, and any sheaf monomorphism τ, we have $\chi_\tau^j(a) = \chi_\tau(a)$.

Proof. The nature of equalisers in SET. $\square$

4. Closed Set Sheaves

Typically sheaves over topological spaces are defined in terms of the open sets of the base space. The notion of a category of sheaves over a site allows us to define a category of sheaves over the closed sets of a topological space and announce that these categories are toposes.

Given a closed set topology τ we can define a covering system C where for each $U \in \tau$ we have $C(U) = \{\{U_i : U_i \in \tau\} : U = \bigcup\{U_i : U_i \in \tau\}\}$. Now, any topology $\mathcal{T}$ is partially ordered by set inclusion, so any topology forms a poset category where all morphisms are the inclusions. For that category let $R = \{U_i \xrightarrow{\alpha_i} U : i \in I\}$ be any pretopology cover. Since between any two objects of $\mathcal{T}$ there can be at most on earrow and it must be an inclusion, we can interpret R to be a family $\{\mathrm{dom}(\alpha_i) : \alpha_i \in R\}$ of elements of $\mathcal{T}$. The defining conditions for a pretopology become:

(i)   for each $U \in \mathcal{T}$, $\{U\} \in P(U)$;

(ii)   if $V \subseteq U$ in $\mathcal{T}$ and $\{U_i : i \in I\} \in P(U)$, then $\{V \cap U_i : i \in I\} \in P(V)$;

(iii)   if $\{U_i : i \in I\} \in P(U)$ and for each $i \in I$ there exists $\{V_{ik}^i : k \in K_i\} \in P(U_i)$, then $\{V_{ik}^i : k \in K_i, i \in I\} \in P(U)$.

Plainly, if we define C as suggested above then, in essence, we have a pretopology for the poset category. Given a pretopology C we can define a (categorial) topology J for category $\mathrm{SET}^{\mathcal{T}^{op}}$ that will give rise to the same category of sheaves.

The constructions for the toposes $sh(\mathcal{T}, J)$ and $sh_j(\mathrm{SET}^{\mathcal{T}^{op}})$ are then standard. We can use either an appropriate version of the equaliser condition Definition 12.2, or Proposition 12.8, to identify those functors in $\mathrm{SET}^{\mathcal{T}^{op}}$ that are sheaves over closed sets. In sum:

Proposition 12.14. The category of sheaves over the closed sets of a topological space forms a topos.

This theory can be extended in a number of directions. One is to identify constructions in toposes of closed set sheaves dual to various morphisms in toposes. It is asserted here that the pseudo-difference arrow (p.108), dual to intuitionist implication, does exist in such categories (proof omitted, see forthcoming).

CHAPTER 13: DUALITY

The first people to take the view that incompleteness and inconsistency are somehow equally reasonable, seem to have been Da Costa and the Brazilian school of logicians, and independently the Routleys. The idea expressed itself in Brazilian logic in the paraconsistent C-logics, dualising by abandoning the law of Noncontradiction $\sim(A\ \&\ \sim A)$, rather than Excluded Middle $A \vee \sim A$ as in intuitionism; and additionally adopting the opposite Double Negation axiom to intuitionism. The Routleys proposed their $*$-operation on theories (see Definition 3.4 or below), which had the property that for any background logic satisfying minimal conditions, the $*$ of an incomplete theory is an inconsistent theory and vice versa. The capacity to admit both inconsistent and incomplete theories was seen as essential to, and explicative of, relevance, at least at the propositional level. Neither the Brazilians nor the Routleys appealed to topological duality, which has only become clear more recently, but which would seem to be an expression of Brazilian intuitions.

This chapter briefly surveys the $*$-operation; then applies it to open and closed set theories, then finds an alternative to the $*$-operation which has different but similar effects.

Recalling Definitions 2.8 and 3.4, an L-semitheory is a set of sentences Th closed under the rule: if A is in Th and $\vdash_L A \to B$ then B is in Th. An L-theory is an L-semitheory closed under conjunctions. An L-semitheory is prime if whenever a disjunction is in it at least one of the disjuncts is also. And for the $*$-operation, $Th^* = \{A : \sim A \text{ is not in } Th\}$.

Proposition 13.1. (Routleys)

(1) If $\vdash_L (A \rightarrow B) \rightarrow (\sim B \rightarrow \sim A)$ then Th is an L-semitheory iff Th^* is an L-semitheory.

(2) If in addition De Morgan's Laws and the Double Negation Law are theorems of L then Th is an L-theory iff Th^* is a prime L-semitheory.

(3) Under the same conditions as (2), Th is a prime L-theory iff Th^* is a prime L-theory.

(4) If $\vdash_L A \leftrightarrow \sim\sim A$ (Double Negation Law) then $Th^{**} = Th$.

(5) If $\vdash_L A \leftrightarrow \sim\sim A$ then Th is inconsistent iff Th^* is incomplete.

(6) If L = classical logic and Th is nontrivial then $Th^* = Th$.

(1), (3) and (6) can be described as *-invariance results. (4) is an involution result. Attention is drawn to (5) which is a duality result of a different kind from the previous chapter. However, * also has interesting effects on theories of open set logics and closed set logics.

Definition 13.2. A theory on a closed set logic or open set logic is *simple* iff sentences are assigned only to either (i) the whole space, (ii) the null set, (iii) any boundary, (iv) any coboundary, that is the whole space minus a boundary.

Obviously (iii) and (iv) cannot both hold if the logic has only open sets, or only closed sets, in it ($PJ4$ has both). Also, theories need not be simple: consider the closed set topology $\{\wedge, \{x\}, (-\infty, x], [x, +\infty), R\}$ where the third and fourth elements are not boundaries.

Proposition 13.3.

(1) Let L be a closed set logic with all elements but F designated. If Th is an inconsistent complete simple theory then Th^* is the consistent complete (simple) theory formed by dropping all sentences A such that both A and $\sim A$ both hold in Th.

(2) Let L be an open set logic with just T designated. If Th is a consistent incomplete simple theory then Th^* is the consistent complete theory formed by adding all sentences A such that neither A nor $\sim A$ hold in Th.

Proof.

(1) Let A and $\sim A$ be in Th. Clearly by the definition of $*$, A is not in Th^*. But in Th, $I(A)$ must be a boundary, so $I(\sim\sim A) =$ the null set, and so $\sim\sim A$ is not in Th. Hence $\sim A$ is in Th^*.

(2) Let neither A nor $\sim A$ be in Th. By the definition of $*$, A is in Th^*. But in Th, $I(A)$ must be a coboundary, so $I(\sim\sim A) =$ the whole space and $\sim\sim A$ is in Th. Hence $\sim A$ is not in Th^*. □

Thus, to inconsistentise a consistent complete theory one can add various atomic sentences A such that $\sim A$ holds, and evaluate as a simple theory in a closed set logic; whereas to incompletise, one can drop various atomic sentences A such that $\sim A$ does not hold and evaluate as a simple theory in an open set logic.

Definition 13.4. $Th^\times = df \{A : \sim A$ is in $Th\}$.

Proposition 13.5.

(1) Let L be a closed set logic and Th be an inconsistent complete simple theory. Then $Th^\times$ is the consistent complete theory formed by dropping all $\sim^{(2n+1)} A$ and adding all $\sim^{(2n)} A$, such that A is atomic and both A and $\sim A$ are in Th.

(2) Let L be an open set logic and Th be a consistent incomplete simple theory. Then $Th^\times$ is the consistent complete theory formed by adding all $\sim^{(2n+1)} A$ and dropping all $\sim^{(2n)} A$, such that A is atomic and neither A nor $\sim A$ are in Th.

Proof.

(1) Let A and $\sim A$ hold in Th. Clearly A is in $Th^\times$. Since Th is simple, $\sim\sim A$ is not in Th, so $\sim A$ is not in $Th^\times$. Since $\sim\sim\sim A$ is in Th, $\sim\sim A$ is in $Th^\times$ etc.

(2) Let neither A nor $\sim A$ be in Th. Clearly A is not in $Th^\times$. Since Th is simple, $\sim\sim A$ is in Th, so $\sim A$ is in $Th^\times$. Since $\sim\sim\sim A$ is not in Th, $\sim\sim A$ is not in $Th^\times$ etc.

□

These results contribute to the duality between incompleteness and inconsistency, and at the same time demonstrate the utility of * and × in connection with these notions.

CHAPTER 14: FOUNDATIONS: PROVABILITY, TRUTH AND SETS
(with Joshua Cole)

1. Introduction

It is appropriate to end with a chapter on topics in what has been called in this century the foundations of mathematics, if only to draw attention to the disavowal of foundationalism in mathematics, but also to draw the attention of mathematicians to the fact that the foundations constitute a mathematically interesting area. Furthermore, the paraconsistent approach grows historically out of logic, which has certainly been part of the usual conception of foundational studies.

Three areas are considered: provability, truth and sets. First there is considered the fate of the classical Gödel sentence and thus the concept of provability, in the finite inconsistent arithmetics. It turns out that it becomes a truth predicate in a certain weak sense. The question of stronger senses of the truth predicate (which can be distinguished inconsistently but not classically) remains open. Second, we review the well-known use of a fixed point method in connection with the truth predicate, as demonstrated by Kripke. Third, we review the application of the fixed point method by Gilmore and Brady to set theory. This demonstrates the existence of incomplete and inconsistent set theories with naive comprehension. In the latter the inconsistent Russell set can be demonstrated to exist (non-well-founded sets). This is a highly desirable state of affairs, since it has the prospect of mathematics being able to rely on the full generality of set abstraction: given any property, one can collect up into a set just the things having that property. It turns out that both uses of the fixed point method produce incomplete as well as inconsistent theories, which are Routley-*-duals of one another.

2. Provability

Gödel's first and second incompleteness theorems arise by means of a partial mapping of the metalanguage of arithmetic into its object language. Specifically, one represents (a) the concept of provability via the provability predicate $\text{Prov}(x)$, and (b) the capacity for self reference via the Diagonal Lemma. From these it is a short argument to the first incompleteness theorem, namely that if an (axiomatisable) arithmetical theory Th is consistent then it is incomplete. On the other hand, if the mapping of the metalanguage were not partial but total in the sense that one were able to represent the truth predicate in the object language while retaining self reference/diagonalisability, then one could demonstrate the inconsistency of Th; that is, Tarski's theorem, which essentially amounts to the Liar paradox. The parallel between the Gödel sentence 'This sentence is unprovable' and the Liar sentence 'This sentence is false', is obvious and striking, and has been noted many times. That is to say, the provability predicate is the nearest one gets within consistent arithmetic to the truth predicate, the Gödel sentence is the consistent arithmetical analog of the Liar sentence, and the first incompleteness theorem is the consistent counterpart of the Liar paradox.

This section aims to contribute to these observations by demonstrating that when one moves to inconsistent extensions of classical arithmetics, specifically the finite moduli arithmetics, then the provability predicate becomes, in a sense to be specified, a truth predicate. The consequence of this result is that in appropriate inconsistent theories the Gödel sentence is, in the same sense, the Liar sentence.

2.1 Consistent Preliminaries

In this section, the terminology and approach to the classical logic case of the Gödel theorems is summarised, drawing on the approach of Boolos and Jeffrey [4a]. We deal with classical arithmetical theories containing Robinson arithmetic Q as a subtheory, including Peano arithmetic $P\#$.

Definition 14.1. If X is any arithmetical sentence, $\ulcorner X \urcorner$ denotes the Gödel number of X. A predicate $F(x)$ is said to be a *provability predicate for a theory* Th iff for any sentence X :

(i)　if $\vdash_{Th} X$ then $\vdash_{Th} F(\ulcorner X \urcorner)$

and (ii)　$\vdash_{Th} F(\ulcorner X \supset Y \urcorner) \supset (F(\ulcorner X \urcorner) \supset F(\ulcorner Y \urcorner))$

and (iii)　$\vdash_{Th} F(\ulcorner X \urcorner) \supset F(\ulcorner F(\ulcorner X \urcorner) \urcorner)$.

If $F(x)$ is a provability predicate for Th, we denote $F(x)$ also by $\mathrm{Prov}_{Th}(x)$; but it should be noted that in general there is more than one provability predicate for a given Th.

Proposition 14.2. Peano Arithmetic $P\#$ and Robinson Arithmetic Q have provability predicates.

Definition 14.3. A set S of sentences is said to be *definable in Q* iff there is a predicate $\mathrm{Prov}_s(x)$ such that for any sentence X

(i)　if $X \in S$ then $\vdash_Q \mathrm{Prov}_s(\ulcorner X \urcorner)$

and (ii)　if $X \notin S$ then $\vdash_Q \sim\mathrm{Prov}_s(\ulcorner X \urcorner)$.

Proposition 14.4. Any recursive set, and in particular the set of theorems of any axiomatisable complete theory, is definable in Q.

Definition 14.5. A function $f(x)$ is said to be *representable in Q* iff there is a predicate $F(x,y)$ such that for any natural numbers a, b if $f(a) = b$ then $\vdash_Q (\forall y)(F(m, y) \equiv y = n)$, where m, n represent a, b respectively in the arithmetical language of F. (This definition generalises to the case where x is an n-tuple but that does not concern us here.)

Proposition 14.6. Every recursive function is representable in Q.

Proposition 14.7. (Diagonal Lemma) For any predicate $F(x)$, there is a sentence H such that $\vdash_Q H \equiv F(\ulcorner H \urcorner)$. In particular, since $\sim\mathrm{Prov}_Q(x)$ and $\sim\mathrm{Prov}_{P\#}(x)$ are such predicates, there are sentences G_Q and $G_{P\#}$ such that $\vdash_Q G_Q \equiv \sim\mathrm{Prov}_Q(\ulcorner G_Q \urcorner)$ and $\vdash_Q G_{P\#} \equiv \sim\mathrm{Prov}_{P\#}(\ulcorner G_{P\#} \urcorner)$.

Intuitively, H "says" in Q and any supertheory: 'This sentence has the property F', while G_Q and $G_{P\#}$ "say" 'This sentence is unprovable in $Q/P\#$'. G_{Th} is called a/the *Gödel sentence for* Th.

Proposition 14.8. Any axiomatisable complete theory containing Q is inconsistent.

Proposition 14.9. (Gödel's first incompleteness theorem). Any consistent axiomatisable theory containing Q is incomplete.

Definition 14.10. A *truth predicate* for a theory Th is any predicate $Tr(x)$ having the property that for any X, $\vdash_{Th} X \equiv Tr(\ulcorner X \urcorner)$.

Proposition 14.11. Any (classical) theory containing Q and having a truth predicate is inconsistent.

2.2 The Inconsistent Case

Moving to the inconsistent finite arithmetics $RM3^i$, we recall the following facts from Chapter 2. The $RM3^i$ are all inconsistent, complete and decidable. More than the last, there is a recursive function $f_R(x)$ returning 2, 1 or 0 respectively as x is the Gödel number of a sentence assigned T, B or F respectively in R, where R is short for any of the $RM3^i$. By representability, therefore, there is an arithmetical predicate F_R with the corresponding representability properties. That is, in Q, $F_R(\ulcorner X \urcorner, y)$ is provably true of 2, 1 or 0 only, where $\ulcorner X \urcorner$ is the Gödel number of a sentence assigned T, B or F respectively in R. It is therefore natural to make the following definition:

Definition 14.12. $\mathrm{Prov}_R(x) =_{df} F_R(x,1) \vee F_R(x,2)$.

Proposition 14.13. The predicate $\mathrm{Prov}_R(x)$ is a provability predicate for R.

Proof. From the previous section, three conditions must be satisfied to be a provability predicate.

Ad (i): If $\vdash_R X$ then by representability (Proposition 14.6), $\vdash_Q F_R(\ulcorner X \urcorner, 2)$. Hence, since $Q \subset R$, $\vdash_R F_R(\ulcorner X \urcorner, 2)$. So, by the three-valued tables for $\vee$, $\vdash_R \text{Prov}_R(\ulcorner X \urcorner)$.

Ad (ii) and (iii): note that Prov_R never takes the value T, for any X. This is because of the transparency of R: back in Q, for fixed $(\ulcorner X \urcorner)$, $F_R(\ulcorner X \urcorner, y)$ is provably true of just one y, namely 2, 1 or 0, and is provably false of all other y. Passing to R, then, the negation of F_R remains provable of all y other than 2, 1 or 0. But in $RM3^3$, say, (that is mod 3) we have that 2, 1 and 0 are provably equal to some of these y (e.g. 5, 4 and 3 respectively). But $RM3^3$ is transparent (as are all the R); that is, provable identities are intersubstitutable in all contexts. So $\sim F_R$ holds of 2, 1, and 0 in R, so that F_R is B or F in R when y is 2, 1 or 0. By the tables for $\vee$, this ensures that for those values of y, Prov_R is B or F also. But by the tables for $RM3$, a $\supset$ sentence with such an antecedent takes a designated value.   □

Now it can be shown that in the above defined sense (Definition 14.10), Prov_R is a truth predicate for R.

Proposition 14.14. For all X, $\vdash_R (X \equiv \text{Prov}_R(\ulcorner X \urcorner))$.

Proof. From the three-valued tables for $\equiv$, $X \equiv Y$ fails to hold only when one side takes the value T and the other side takes the value F. Now if X is T then since Prov is a provability predicate, then by (i) of the previous proposition, $\vdash_R \text{Prov}_R(\ulcorner X \urcorner)$. That is, the latter does not take the value F in R. On the other hand, by the argument for (ii) and (iii) of the previous Proposition, $\text{Prov}_R(\ulcorner X \urcorner)$ is never T. Consequently, when X is F, their $\equiv$ holds as required.   □

This proposition is not a trivial reconstruction by means of the Extendability lemma of something which holds in Q: $\vdash_R$ cannot be replaced by $\vdash_Q$ else Q would be inconsistent (by the Diagonal lemma applied to the predicate $\sim \text{Prov}_R(x)$). It is now shown that the Gödel sentence models the Liar sentence in the sense that both the Gödel sentence and its negation are theorems of the $RM3^i$. The Gödel sentence for R is that sentence whose existence is guaranteed by diagonalisation in Q; that is, the sentence which expresses in Q the statement 'I am unprovable in R'.

Proposition 14.15. $\vdash_R G_R$ and $\vdash_R \sim G_R$.

Proof. First note that the existence of G_R is ensured by the classical Diagonal Lemma (Proposition 14.7), as the sentence such that given the predicate $\sim\text{Prov}_R(x)$, $\vdash_Q G_R \equiv \sim \text{Prov}_R(\ulcorner G_R \urcorner)$. Suppose not $\vdash_R G_R$. By representability, $\vdash_Q \sim \text{Prov}_R(\ulcorner G_R \urcorner)$. By the Diagonal Lemma, $\vdash_Q G_R$, so that, since $Q \subset R$, $\vdash_R G_R$. That is, from not $\vdash_R G_R$ it follows that $\vdash_R G_R$; hence $\vdash_R G_R$. By representability again, $\vdash_Q \text{Prov}_R(\ulcorner G_R \urcorner)$. By the Diagonal Lemma again, $\vdash_Q \sim G_R$. But $Q \subset R$, so $\vdash_R \sim G_R$.   □

A further interesting fact follows here: that the denial of the Gödel sentence for any of the R is already provable in Q. This is as it should be: since the R are decidable, when G is provable in R then this fact is provable in Q, hence by diagonalisation so is the denial of G. It further follows that if Q is consistent, then the Gödel sentence for R is unprovable in Q.

Proposition 14.16. (Strong diagonalisation for G_R).

$$\vdash_R G_R \leftrightarrow \sim \text{Prov}_R(\ulcorner G_R \urcorner).$$

Proof. From Proposition 14.14, G_R is B in any of the R. Since Prov_R is a provability predicate and G_R holds, so does Prov_R. But by the argument of the previous proposition Prov_R is never T. Thus Prov_R is exactly B, so that $\sim\text{Prov}_R$ is also. The $RM3$ tables for $\leftrightarrow$ then ensure the Proposition.   □

Problem: Prov is a weak truth predicate in the sense that the $\equiv$ of Proposition 14.4 does not guarantee detachability in general; though being a provability predicate there is detachability one way (Proposition 14.13(i)). We have just seen the stronger detachability is present for G_R via the stronger connective $\leftrightarrow$. Can Proposition 14.14 be reproved with $\leftrightarrow$ replacing $\equiv$, or at least detachability the other way, perhaps with a modification to the natural Prov that we have used?

There is a point here about Löb's theorem. This says that, in any theory Th extending Q and in the same language, if F is a provability predicate for Th, then if $\vdash_{Th} F(\ulcorner A \urcorner) \supset A$ then $\vdash_{Th} A$. This might look like an insuperable barrier to Prov being a truth predicate, since a truth predicate would have the above properties so that triviality would follow. However, Löb's theorem breaks down in the inconsistent finite models. This is because the proof of Löb's theorem essentially requires one to deduce the consequent of a $\supset$ statement from its antecedent, which one cannot do in these inconsistent nontrivial theories. Now it is possible that a stronger version of Löb's theorem can be obtained for the stronger language containing $\to$ as well as $\supset$, since the property $\vdash Tr(\ulcorner A \urcorner) \leftrightarrow A$ would certainly ensure that Tr would be a provability predicate, and the $\to$ has the necessary deductive force. However, while possible, this is not obviously true, since adding to the language formally voids the diagonal lemma which is applied in the usual proof of Löb's theorem, so that Löb's theorem might break down for this addition to the language. In the following sections it is seen that a truth predicate can be added to the classical language without triviality.

3. Truth

3.1 The Fixed Point Method

The fixed point method is an iterative method for constructing a model for a collection of axioms. In general terms we take as starting point an already established model and extend it by adding new logical predicate symbols with their attendant governing axioms. The axioms to be modelled need to have a conditional or biconditional main connective and may be quantified. The model for the new axioms is given a (transfinite) inductive definition. Some simple rules are iterated to eventually produce the new extended model. It should be noted that the underlying logic of models generated in this way are non-classical. Depending on your philosophical disposition the logic can be incomplete or inconsistent, as we shall see.

These sections will be structured in the following way. Firstly the method is described in general terms. Then, contrary to historical sequence, it is shown how the method has been used by Kripke to model the T-scheme in languages which contain their own truth predicate. Then we see how the method can be used to model axioms in set theory. In particular it provides a method for modelling the axioms of comprehension and extensionality. Gilmore appears to have been the first to use the method to model the axiom of comprehension in a restricted, non-extensional set theory. Later Brady showed how to model the extensionality axiom as well in an inconsistent logic. Then followed Kripke, while later use of the method has been made by Feferman in connection with the foundations of category theory. Finally there is brief speculation on the possibility of further applications for this method.

Suppose that we have a language $\mathcal{L}$ which we wish to extend by adding an n-ary predicate P^n. Let $\mathcal{L}$ have an interpretation I with domain D. The usual interpretation for a predicate is in terms of a subset of D^n. That is, an n-ary relation on D. We define the interpretation of P^n in terms of an ordered pair (S_1, S_2) where both S_1 and S_2 are n-ary relations on D. S_1 is called the *extension of P* and S_2 is called the *antiextension*. The axiom to be modelled will be of a biconditional form: A iff B, where A is a sentence of the form $P^n(a_1, a_2, \ldots, a_n)$ and B is a formula containing some of $a_1, a_2, \ldots, a_n$.

Basically the model is built up in stages by repeatedly adding to the interpretation of P to force sentences which are instances of the axiom's LHS A to be interpreted true or false whenever its corresponding RHS B is interpreted as true or false respectively at the previous stage. Each time the model is extended in this way a whole new collection of sentences become available as true RHSs of the axiom, thus requiring the model to be again expanded to include a new collection of corresponding LHSs to acount for the axiom's truth.

Eventually, through the magic of infinity, there will be a stage where for each candidate RHS true (false) in the interpretation, its corresponding LHS will already be interpreted true (false). Such points will be called fixed points. So we have extended the original language $\mathcal{L}$ to contain a new predicate P^n. We have extended the original interpretation I by interpreting P^n in terms of the pair (S_1, S_2). The interpretation of everything else remains unchanged. Let I_0 be the original interpretation I. The fixed point method will generate a succession of interpretations $I_0, I_1, I_2, \ldots$. In successive interpretations the interpretaiton of P^n which is the ordered pair (S_1, S_2) is modified by extending the relations of S_1, S_2 or both.

If we think of successive changes to the interpretation as being the results of applications of a rule ϕ, i.e. $I_{n+1} = \phi(I_n)$, any fixed point of ϕ, i.e. interpretation I_λ such that $\phi(I_\lambda) = I_\lambda$, will be an interpretation which will model the extended language.

3.2 The fixed point method applied to truth theory

The liar paradox and its variations have been thought to arise from the capacity of languages to express their own truth and falsity predicates. The Tarski hierarchy of languages is an elaborate attempt to avoid the paradoxes by postulating infinitely many levels of language, each with its own truth predicate $True_n$. Sentences at level n can only be named in level $n+1$ and greater. This is complicated and unnatural and doesn't correspond well with the facts. In natural languages we are clearly able to name sentences in the language without jumping to higher and higher levels. There is only one level not many.

In his 1975 paper 'Outline of a Theory of Truth' [19], Kripke showed that it is possible for a language to contain its own truth predicate and yet still avoid the liar and related paradoxes. Using the fixed point method Kripke outlined a more intuitive theory of truth than the Tarski language hierarchy. Although Kripke was not the first to use the method, his paper is used to illustrate the fixed point method because its application to truth theory has particular intuitive appeal.

The truth predicate T is governed by the T-scheme axiom:

$$T(\ulcorner A \urcorner) \equiv A$$

Using the fixed point method we are able to show that an uncontroversial language not containing its own truth predicate can be extended to contain its own truth predicate which behaves according to the T-scheme. We start with a language $\mathcal{L}$ $(\neg, \vee, \&, \supset, \equiv, \exists, \forall)$ and extend it by adding the unary predicate $T(x)$. The axiom to be modelled, the T-scheme, has a $\equiv$ main connective.

We will interpret the truth predicate T in the extended language by the pair (S_1, S_2) of unary relations on D (i.e. subsets of D). The interpretation of every-thing else in the extended language will remain as before, closed sentences being interpreted in the values $\{True, False\}$. Initially T will be interpreted as $(\emptyset, \emptyset)$ and it will be built up in stages by applying a function ϕ until a fixed point is reached.

The function ϕ appends new sentences to the extension S_1 and antiextension S_2 of T according to the wff evaluation rules. That is $\phi((S_1, S_2)) = (S_1', S_2')$. The wff evaluation rules are applied with T only partially interpreted by (S_1, S_2) and the elements of D which are codes of sentences interpreted $True$ are collected together to form S_1'. Elements of D which are not codes of sentences or are the codes of sentences intepreted $False$ are collected together to form S_2'. This explains how interpretations $I_0, I_1, I_2, \ldots$ are defined. For a limit ordinal the situation is different. Let λ be a limit ordinal, then $S_{i,\lambda} = \cup_{\alpha < \lambda} S_{i,\alpha}$ for $i = 1, 2$.

The problem can be seen as the problem of explaining to someone the notion of truth. We assume that they understand the meanings of all sentences in the language except those containing the word 'true'. The initial complete ignorance of the notion of truth is indicated by the empty interpretation $(\emptyset, \emptyset)$. The concept of truth is built up in stages by applying a simple rule: we are entitled to assert (or deny) that any sentence is true (or false) under the exact same circumstances we can assert (or deny) that sentence itself. This is the function ϕ in our formalisation. Our subject has a complete understanding of when sentences not containing the word 'true' can be asserted or denied. By applying the rule about truth they are able to glean a partial understanding of sentences containing the word 'true'. By applying the rule once from a situation of complete ignorance sentences like

" 'Dogs are mammals' is true" and " 'The moon is a piece of cheese' is true" will be interpreted as *True* and *False* respectively. However, sentences like ' " 'Dogs are mammals' is true" is true' will still not be interpreted. This partial understanding of sentences containing the word 'true' is formalised by $(S_{1,1}, S_{2,1})$. Applying the rule governing truth again will result in this last sentence being correctly interpreted *True*. At a fixed point I_λ, $\phi((S_{1,\lambda}, S_{2,\lambda})) = (S_{1,\lambda}, S_{2,\lambda})$. So, applying the rule about the word 'true' adds nothing further to the concept. At such a point we will say that the model is saturated and the language contains its own truth predicate.

Now the theory of truth which results from this treatment can be shown to be incomplete. That is, some sentences never get assigned a truth value. Some sentences are considered neither true nor false. An example is the Liar sentence. Assume it does get assigned a truth value and let α be the least ordinal such that $I_\alpha(\text{LIAR}) = \textit{True(False)}$. Once we permit such a sentence to have a truth value we unleash its paradoxical potential. Applying the rule ϕ we generate $I_{\alpha+1}$ which has the effect of reversing the Liar's truth value. That is $I_{\alpha+1}(\text{LIAR}) = \textit{False(True)}$. This contradicts the monotonicity of the operation ϕ which will be proved in the section which follows.

3.3 The proof that fixed points model the T-scheme

To prove that fixed points provide a model for the T-scheme axiom we require a simple lemma about the construction.

Lemma 1. If $\alpha \leq \beta$ then for any sentence S if $I_\alpha(S) = \textit{True}$, then $I_\beta(S) = \textit{True}$. If $\alpha(S) = \textit{False}$, then $I_\beta(S) = \textit{False}$.

This lemma says that the interpretation of the truth predicate T is only ever changed by giving a truth value to sentences which were previously not interpreted. Once a sentence becomes interpreted as a truth value, its value never changes in subsequent interpretations. If we define the relation $\leq$ between interpretations as $I_\alpha \leq I_\beta$ iff $S_{1,\alpha} \subseteq S_{1,\beta}$ and $S_{2,\alpha} \subseteq S_{2,\beta}$ then lemma 1 says that ϕ is a monotone increasing operation on $\leq$.

Proof of Lemma 1. The proof is by induction on the number of connectives in S. If S is an atomic sentence then lemma 1 holds. Assume for sentences A and B that the lemma holds. We prove that lemma 1 holds for a sentence S where S is of the form $\neg A, A \vee B, A\&B, (\exists x)A, (x)A$.

e.g. Let $I_\alpha(A\&B) = \textit{True}$. By the wff evaluation rules $I_\alpha(A) = I_\alpha(B) = \textit{True}$. By the induction hypothesis, $I_\beta(A) = I_\beta(B) = \textit{True}$. So $I_\beta(A\&B) = \textit{True}$. There is a similar proof for $I_\alpha(A\&B) = \textit{False}$.

Divide the proof that the fixed point method generates a model for the T-scheme into two parts: one for each direction of the biconditional. The modelling of the schema for general sentence α is proved by showing the modelling of one instance of the schema for an arbitrarily chosen sentence S.

Let I_λ be a fixed point of the method. That is, $\phi((S_{1,\lambda}, S_{2,\lambda})) = (S_{1,\lambda}, S_{2,\lambda})$.

Left to right: We assume $I_\lambda(T(S)) = \textit{True}$, and show that $I_\lambda(S) = \textit{True}$. Let α be the least ordinal for which $I_\alpha(T(S)) = \textit{True}$. α will be a successor ordinal because (i) it is non-zero, and (ii) if it were a limit ordinal then by the method of construction there would be an ordinal $\beta < \alpha$ such that $I_\beta(T(S)) = \textit{True}$, thus contradicting the assumption that α was the least such ordinal.

If α is a successor ordinal and the least ordinal for which $I_\alpha(T(S)) = \textit{True}$, then by the method of construction it must be the case that $I_{\alpha-1}(S) = \textit{True}$. Now, since $I_{\alpha-1} \leq I_\alpha$ it follows by lemma 1 that $I_\alpha(S) = \textit{True}$ as required. A similar proof can be run assuming $I_\lambda(T(S)) = \textit{False}$ and showing that $I_\lambda(S) = \textit{False}$.

Right to left: We assume $I_\lambda(S) = \textit{True}$ and show that $I_\lambda(T(S)) = \textit{True}$. By the method of construction it follows that $I_{\lambda+1}(T(S)) = \textit{True}$. But since I_λ is a fixed point $I_\lambda = I_{\lambda+1}$. So $I_\lambda(T(S)) = \textit{True}$ as required. If $I_\lambda(S) = \textit{False}$ it follows by a similar argument that $I_\lambda(T(S)) = \textit{False}$. $\square$

3.4 The proof that there are fixed points

The method has a certain intuitive appeal and fixed points would seem to be the kind of solution to the problem of representing the truth predicate that we are looking for. However, it is not immediately obvious that fixed point solutions will always exist.

The proof that there are fixed points is a simple argument based on some assumptions about the language being modelled and the function ϕ. The method will generate a chain of ordered pairs of n-ary relations on D modelling the predicate P^n in our language:

$$(S_{1,0}, S_{2,0}) \leq (S_{1,1}, S_{2,1}) \leq (S_{1,2}, S_{2,2}) \leq \ldots$$

At each stage in the construction more sentences in the language (or their names) are being added to the extension and antiextension of P^n. At every stage, at least one new sentence gets decided. An assignment to extension or antiextension is never changed by a later assignment. The relations of extension and antiextension only increase in size, they never retract. This feature of the method results from the monotonicity of the function ϕ.

There are only denumerably many sentences our language which contain the predicate P^n. You could show a 1-1 correspondence with the natural numbers by listing the sentences containing P^n in alphabetical order or similar.

So for some λ of the second number class, $(S_{1,\lambda}, S_{2,\lambda}) = \phi((S_{1,\lambda}, S_{2,\lambda}))$.

4. The Fixed Point Method Applied to Set Theory

P.C. Gilmore appears to have been the first person to use the fixed point method in a recognisable form although he attributes its origins to a persistence lemma by Roger C. Lyndon in 1959. In his paper, 'The Consistency of Partial Set Theory Without Extensionality' [12a], Gilmore shows how to model the comprehension axiom in a partial set theory. To say that the theory is partial means that for a set S in the theory, its characteristic function

$$f_S(x) = \begin{cases} 1 & \text{if } x \in S \\ 0 & \text{if } x \notin S \end{cases}$$

is a partial function.

The theory extends predicate logic with two new primitive binary predicates $\in$ and $\notin$. New terms in the theory are formulae of the form $\{x : P, Q\}$ where P and Q are 'positive' formulae. That is, formulae in which only conjunction, disjunction and quantifications are used.

The axioms to be modelled are the pair:

(1) $(x)[(x \in \{y : P(y), Q(y)\} \vee (P(x)\&Q(x))) \equiv P(x)]$

(2) $(x)[(x \notin \{y : P(y), Q(y)\} \vee (P(x)\&Q(x))) \equiv Q(x)]$

The main connectives in these two axioms are disjunctions and the method requires a conditional or biconditional main connective. Gilmore gives a pair of conditional sentences for each axiom (1) and (2) whose conjunction is logically equivalent to the original axiom.

(1.1) $(x)[(x \in \{x : P(x), Q(x)\}) \supset P(x)]$

(1.2) $(x)[(P(x)\&\neg Q(x)) \supset x \in \{y : P(y), Q(y)\}]$

(2.1) $(x)[(x \notin \{x : P(x), Q(x)\}) \supset Q(x)]$

(2.2) $(x)[\neg(P(x)\&Q(x)) \supset x \notin \{y : P(y), Q(y)\}]$

He shows how these four sentences are modelled using the fixed point method.

We interpret $\in$ and $\notin$ by a pair of sets (S_1, S_2). S_1 and S_2 are built up in stages by applying a simple rule: assume we have $(S_{1,\alpha}, S_{2,\alpha})$ already defined for an ordinal α. We generate $S_{1,\alpha+1}$ by appending to $S_{1,\alpha}$ sentences $a \in \{x : P(x), Q(x)\}$ for all sentences $P(a)\&\neg Q(a)$ such that $I_\alpha(P(a)\&\neg Q(a)) = True$. Similarly we generate $S_{2,\alpha+1}$ by appending to $S_{2,\alpha}$ sentences $a \notin \{x : P(x), Q(x)\}$ for all sentences $\neg P(a)\&Q(a)$ such that $I_\alpha(\neg P(a)\&Q(a)) = True$. This amounts to a definition of the function ϕ.

For a limit ordinal λ, $S_{i,\lambda} = \cup_{\alpha < \lambda} S_{i,\alpha}$ for $i = 1, 2$.

We let I_0 of the construction be a model for the language without the new predicates $\in$ and $\notin$ with $(S_{1,0}, S_{2,0}) = (\emptyset, \emptyset)$.

Again the proof that the fixed point method generates a model for the set theory axioms requires first a lemma that ϕ is monotone.

Lemma 2. If $\alpha \leq \beta$ then for any sentence S if $I_\alpha(S) = \text{True}$, then $I_\beta(S) = \text{True}$. If $I_\alpha(S) = \text{False}$, then $I_\beta(S) = \text{False}$ (cf. Lemma 1).

Proof. The proof proceeds as an induction on the number of logical connectives in the sentence S. This is Lyndon's persistence lemma. □

Armed with this lemma we can now set about proving that the fixed point method generates a model for the four sentences $(1.1), (1.2), (2.1), (2.2)$. Let I_λ be a fixed point. That is, let $I_\lambda = I_{\lambda+1}$.

(1.1): Assume for arbitrary a that $I_\lambda(a \in \{x : P(x), Q(x)\}) = \text{True}$. We want to show that $I_\lambda(P(a)) = \text{True}$. Let α be the least ordinal for which $I_\alpha(a \in \{x : P(x), Q(x)\}) = \text{True}$. α is a successor ordinal. So by the method of construction $I_{\alpha-1}(P(a)\&\neg Q(a)) = \text{True}$. Now $\alpha - 1 \leq \lambda$, so by Lemma 1 $I_\lambda(P(a)\&\neg Q(a)) = \text{True}$. Hence $I_\lambda(P(a)) = \text{True}$.

(2.1): Assume for arbitrary a that $I_\lambda(a \notin \{x : P(x), Q(x)\}) = \text{True}$. We want to show that $I_\lambda(Q(a)) = \text{True}$. Let α be the least ordinal for which $I_\alpha(a \in \{x : P(x), Q(x)\}) = \text{True}$. α is a successor ordinal. So by the method of construction $I_{\alpha-1}(\neg P(a)\&Q(a)) = \text{True}$. Now $\alpha - 1 \leq \lambda$, so by Lemma 1, $I_\lambda(\neg P(a) \& Q(a)) = \text{True}$. Hence $I_\lambda(Q(a)) = \text{True}$.

(1.2): Assume for arbitrary a that $I_\lambda(P(a)\&\neg(Q(a)) = \text{True}$. We want to show that $I_\lambda(a \in \{x : P(x), Q(x)\}) = \text{True}$. By the method of construction $I_{\lambda+1}(a \in \{x : P(x), Q(x)\}) = \text{True}$. But $I_\lambda = I_{\lambda+1}$, so $I_\lambda(a \in \{x : P(x), Q(x)\}) = \text{True}$.

(2.2): Assume for arbitrary a such that $I_\lambda(\neg P(a) \& Q(a)) = True$. We want to show that $I_\lambda(a \notin \{x : P(x), Q(x)\}) = True$. By the method of construction $I_{\lambda+1}(a \notin \{x : P(x), Q(x)\}) = True$. But $I_\lambda = I_{\lambda+1}$, so $I_\lambda(a \notin \{x : P(x), Q(x)\}) = True$. $\qquad\square$

We can be assured of the existence of fixed points by virtue of a proof similar to last time.

Like Kripke's truth theory, Gilmore's set theory is incomplete. Neither of the following sentences will be interpreted true:

$$R' \in R'$$
$$R' \notin R'$$

where $R' = \{x : x \in x, x \notin x\}$.

Let α be the first stage at which $I_\alpha(R' \in R') = True$. By the method of construction α is a successor ordinal and $I_{\alpha-1}(R' \in R') = True$ contradicting the assumption that α was the least such ordinal. A similar proof can be provided for $R' \notin R'$.

So we are able to model a version of the axiom of comprehension of set theory using this method. Gilmore's paper is significant because it led the way for a number of useful applications of the method. In 1971 Ross Brady showed that both the axiom of comprehension and a version of the axiom of extensionality could be modelled using transfinite induction in a paper 'The Consistency of the Axioms of Abstraction and Extensionality in a Three-Valued Logic' [5a]. This paper is interesting not so much for its modelling of extensionality, but because it models the axioms in paraconsistent three-valued logic. Where Kripke and Gilmore are incomplete, Brady is inconsistent. But the difference between the modellings is only superficial – inconsistent models can be simply transformed to become incomplete and vice versa.

Brady models a more familiar version of the axiom schema of comprehension:

$$(\exists y)(x)(x \in y \leftrightarrow P(x, z_1, \ldots, z_n))$$

We extend propositional logic by adding the new primitive predicate symbol $\in$ which is governed by a familiar comprehension axiom. Brady's model has a background logic Lukasiewicz three-valued. The values are $1, \frac{1}{2}, 0$ and the first two are designated.

Initially all wffs of the form $x \in y$ are assigned the truth value $\frac{1}{2}$. This is the initial interpretation I_0. As the construction proceeds, more and more of these sentences get assigned either a value 0 or 1 according to a rule ϕ : assuming we have I_α defined for some ordinal α. Then $I_{\alpha+1}$ is generated from I_α by making the following changes: $I_{\alpha+1}(a \in \{x : P(x)\}) = I_\alpha(P(a))$. That is, sentences $a \in \{x : P(x)\}$ get assigned a value 1 (0) at stage $\alpha+1$ whenever the sentence $P(a)$ was interpreted 1 (0) at stage α. For a limit ordinal λ, $I_\lambda(a \in \{x : P(x)\}) = 1$ (0) if for some $\alpha \leq \lambda, I_\alpha(P(a)) = 1$ (0).

Like in the previous two constructions, we again have a set of sentences which is the extension of the new predicate symbol and a set of sentences which is its antiextension. These are the sentences assigned the values 1 and 0 respectively. Unlike the previous constructions every sentence gets assigned a truth value, because initially all wffs of the form $x \in y$ are assigned the value $\frac{1}{2}$. Now $\frac{1}{2}$ is a designated truth value so that paradoxical sentences involving $\in$ get assigned a designated truth value.

It can be proved that $R \notin R$ where $R = \{x : x \notin x\}$ never gets assigned a value 0 or 1 because that would contradict the monotonicity of ϕ. Also, the sentence $R' \in R'$ where $R' = \{x : x \in x\}$ never gets assigned a truth value 0 or 1 because there can be no first level at which this value is assigned. Such sentences remain with the initial truth value $\frac{1}{2}$. But it is only a matter of convention whether sentences which never get assigned a truth value of true or false by the method are considered to have no truth value at all, or a third undesignated truth value, or a third designated truth value. The method does not stipulate which.

The axiom of extensionality for set theory does not lend itself to a simple modelling by the fixed point method. This is because it is not in required form. We add to our language a new predicate = for sets and model the axiom:

$$(x)(y)(z)(x = y \supset (x \in z \equiv y \in z))$$

This has a conditional main connective but we need the sentence containing the new predicate = on its right for the method to be simply applied.

Brady models a different axiom of extensionality:

$$(x)(y)[(v)(v \in x \leftrightarrow v \in y) \supset (z)(x \in z \leftrightarrow y \in z)]$$

using transfinite induction. This axiom does not contain any new predicate symbol = and the construction does not fit the general fixed point method procedure.

5. Further Applications

Clearly the fixed point method has quite general application in mathematics. It can be used to model axioms which introduce a new predicate symbol and which are in the appropriate form. The purpose of studying this method was in the hope that it could be used to model some axioms of category theory.

The current foundations for category theory seem unnecessarily restrictive in the kinds of categories they allow us to construct. Categories are restricted in size so as to avoid Russell-type paradoxes. The thought is that it might be possible to give a comprehension axiom for category theory which permits the construction of the types of categories which seem intuitively possible but which are forbidden by the current foundations. The modelling of such an axiom could perhaps employ methods similar to ones used in here.

BIBLIOGRAPHY

[1] Anderson, Alan and Nuel Belnap, *Entailment*, Princeton, 1975.

[2] Asenjo, F., 'Towards an Antinomic Mathematics' in [48].

[3] Bell, J., 'Infinitesimals', *Synthese*, **75** (1988), 285-316.

[4] Birkhoff, G. and S. MacLane, *A Survey of Modern Algebra*, 3rd edition, New York, Macmillan, 1965.

[4a] Boolos, G. and R.C. Jeffrey, *Computability and Logic* (Second Edition), Cambridge, Cambridge University Press, 1980.

[5] Brady, Ross, 'The Nontriviality of Dialectical Set Theory', in [48].

[5a] Brady, Ross, 'The Consistency of the Axioms of Abstraction and Extensionality in a Three-Valued Logic', *Notre Dame Journal of Formal Logic (NDJFL)*, **12** (1971), 447-453.

[6] Burgess, J., 'Relevance: a Fallacy?', *NDJFL*, **22** (1981), 97-104.

[7] Burgess, J., 'Common Sense and Relevance', *NDJFL*, **24** (1983), 41-53.

[8] Coleman, Edwin, *The Role of Notation in Mathematics*, Ph.D. dissertation, The University of Adelaide, 1988.

[9] Coxeter, H.S.M., *Introduction to Geometry*, New York, Wiley, 1967.

[10] Da Costa, Newton, C.A., 'On the Theory of Inconsistent Formal Systems', *NDJFL*, **15** (1974), 497-510.

[11] Dunn, J. Michael, 'A Theorem in Three Valued Model Theory with Connections to Number Theory, Type Theory, and Relevant Logic', *Studia Logica*, **38** (1979), 149-169.

[12]   Fitting, Melvin, 'Bilattices and the Theory of Truth', *Journal of Philosophical Logic*, **18** (1989), 225-256.

[12a]   Gilmore, P.C., 'The Consistency of Partial Set Theory Without Extensionality', in D. Scott (ed.), Symposium in Pure Mathematics, University of California, Los Angeles 1967: Axiomatic Set Theory, *Proceedings of Symposia in Pure Mathematics*, Vol.XIII Pt.2, AMS 1974.

[13]   Goodman, Nicholas, 'The Logic of Contradictions', *Zeitschrift fur Mathematische Logic und Grundlagen der Mathematik*, **27** (1981), 119-126.

[14]   Goldblatt, Robert, *Topoi*, Revised edn., Amsterdam, North-Holland, Elsevier, 1984.

[15]   Hatcher, W.S., *The Logical Foundations of Mathematics*, Oxford Pergamon, 1982.

[15a]   Johnstone, P.T., *Topos Theory*, Academic Press, 1977.

[16]   Kelley, J.L., *General Topology*, New York, Van Nostrand, 1955.

[17]   Keisler, Jerome, *Foundations of Infinitesimal Calculus, and Elementary Calculus*, Boston, Prindle, Weber and Schmidt, 1976.

[18]   Kock, Anders, *Synthetic Differential Geometry*, Cambridge University Press, 1981.

[19]   Kripke, Saul, 'Outline of a Theory of Truth', *The Journal of Philosophy*, **72** (1975), 690-716.

[20]   Lavers, Peter, 'Relevance and Disjunctive Syllogism', *NDJFL*, **29** (1988), 34-44.

[21]   Lawvere, F.W., and S.H. Schanuel, *Categories in Continuum Physics*, Springer Lecture Notes in Mathematics No.1174, 1986.

[22]   Mendelson, E., *Introduction to Mathematical Logic*, New York, Van Nostrand Reinhold, 1964.

[23] Meyer, Robert K., 'Relevant Arithmetic', *Bulletin of the Section of Logic of the Polish Academy of Science*, **5** (1976), 133-137.

[24] Meyer, Robert K., *The Consistency of Arithmetic, Research Papers in Logic*, Australian National University, 1976.

[25] Meyer, Robert K., *Arithmetic Formulated Relevantly, Research Papers in Logic*, Australian National University, 1976.

[26] Meyer, Robert K. and Errol Martin, 'Logic on the Australian Plan', *The Journal of Philosophical Logic*, **15** (1986), 305-332.

[27] Meyer, Robert K. and Chris Mortensen, 'Inconsistent Models for Relevant Arithmetics', *The Journal of Symbolic Logic*, **49** (1984), 917-929.

[28] Meyer, Robert K. and Chris Mortensen, *Alien Intruders in Relevant Arithmetic, Technical Reports in Automated Reasoning*, TR-ARP-6/87 (1987).

[29] Meyer, Robert K., R. Routley and J.M. Dunn, 'Curry's Paradox', *Analysis*, **39** (1979), 124-128.

[30] Mortensen, Chris, 'Every Quotient Algebra for C_1 is Trivial', *NDJFL*, **21** (1980), 694-700.

[31] Mortensen, Chris, 'The Validity of Disjunctive Syllogism is Not So Easily Proved', *NDJFL*, **24** (1983), 35-40.

[32] Mortensen, Chris, 'Reply to Burgess and to Read', *NDJFL*, **27** (1986), 319-337.

[33] Mortensen, Chris, 'Inconsistent Nonstandard Arithmetic', *The Journal of Symbolic Logic*, **52** (1987), 512-518.

[34] Mortensen, Chris, 'Inconsistent Number Systems', *NDJFL*, **29** (1988), 45-60.

[35] Mortensen, Chris, 'Anything is Possible', *Erkenntnis*, **30** (1989), 319-337.

[36] Mortensen, Chris, 'Paraconsistency and C_1', in [48].

[37] Mortensen, Chris, 'Models for Inconsistent and Incomplete Differential Calculus', *NDJFL*, **31** (1990), 274-285.

[38] Mortensen, Chris, and Tim Burgess, 'On Logical Strength and Weakness', *History and Philosophy of Logic*, **10** (1989), 47-51.

[39] Mortensen, Chris, and Peter Lavers, '$\mathcal{E}$-semantics for Dual Toposes is paraconsistent', (manuscript).

[40] Mortensen Chris, and Steve Leishman, *Computing Dual Paraconsistent and Intuitionist Logics, Technical Reports in Automated Reasoning*, TR-ARP-9/89, 1989.

[41] Mortensen Chris, and Robert K. Meyer, 'Relevant Quantum Arithmetic', in de Alcantara (ed.) *Mathematical Logic and Formal Systems*, New York, Marcel Dekker, 1984, 221-226.

[42] Mortensen Chris and Graham Priest, 'The Truth Teller Paradox', *Logique et Analyse*, **95-6** (1981), 381-388.

[43] Patterson, E.M., *Topology*, (Second Edition), Edinburgh, Oliver and Boyd, 1963.

[44] Perlis, S., *Theory of Matrices*, Reading, Addison-Wesley, 1952.

[45] Priest, Graham, 'The Logic of Paradox', *The Journal of Philosophical Logic*, **8** (1979), 219-241.

[46] Priest, Graham, *In Contradiction*, Dordrecht, Nijhoff, 1987.

[47] Priest, Graham and Richard Routley, *On Paraconsistency*, Research Papers in Logic 13, A.N.U., 1983, reprinted in [48].

[48] Priest, Graham, Richard Routley and Jean Norman (eds.), *Paraconsistent Logic*, Munchen, Philosophia Verlag, 1989.

[49] Read, Stephen, 'Burgess on Relevance: a Fallacy Indeed', *NDJFL*, **24** (1983), 473-481.

[50] Rogers, R., *Mathematical Logic and Formalised Theories*, Amsterdam, North-Holland, 1971.

[51] Routley, Richard, *Ultralogic as Universal*, reprinted as the appendix in his *Exploring Meinong's Jungle and Beyond*, Australian National University, 1980.

[52] Routley, Richard and Val Routley, 'The Semantics of First Degree Entailment', *Nous*, **6** (1972), 335-359.

[53] Slaney, John, 'RWX is not Curry Paraconsistent', in [48].

[54] Stroyan, K., 'Infinitesimal Analysis of Curves and Surfaces', in J. Barwise (ed.), *Handbook of Mathematical Logic*, Amsterdam, North-Holland, 1977.

[55] Tall, David, 'Looking at Graphs Through Infinitesimal Microscopes, Windows, and Telescopes', *The Mathematical Gazette*, **64** (1980), 22-49.

[56] Tarski, Alfred, *A Decision Method for Elementary Algebra and Geometry*, Berkeley, University of California Press, 1981.

INDEX OF DEFINITIONS & NAMES

Other *Mathematics and Its Applications* titles of interest:

P.H. Sellers: *Combinatorial Complexes. A Mathematical Theory of Algorithms.*
1979, 200 pp.   ISBN 90-277-1000-7

P.M. Cohn: *Universal Algebra.* 1981, 432 pp.
                    ISBN 90-277-1213- 1 (hb), ISBN 90-277-1254-9 (pb,

J. Mockor: *Groups of Divisibility.* 1983, 192 pp.   ISBN 90-277-1539-4

A. Wwarynczyk: *Group Representations and Special Functions.* 1986, 704 pp.
                    ISBN 90-277-2294-3 (pb), ISBN 90-277-1269-7 (hb)

I. Bucur: *Selected Topics in Algebra and its Interrelations with Logic, Number
Theory and Algebraic Geometry.* 1984, 416 pp.   ISBN 90-277-1671-4

H. Walther: *Ten Applications of Graph Theory.* 1985, 264 pp.
                                                    ISBN 90-277-1599-8

L. Beran: *Orthomodular Lattices. Algebraic Approach.* 1985, 416 pp.
                                                    ISBN 90-277-1715-X

A. Pazman: *Foundations of Optimum Experimental Design.* 1986, 248 pp.
                                                    ISBN 90-277-1865-2

K. Wagner and G. Wechsung: *Computational Complexity.* 1986, 552 pp.
                                                    ISBN 90-277-2146-7

A.N. Philippou, G.E. Bergum and A.F. Horodam (eds.): *Fibonacci Numbers and
Their Applications.* 1986, 328 pp.   ISBN 90-277-2234-X

C. Nastasescu and F. van Oystaeyen: *Dimensions of Ring Theory.* 1987, 372 pp.
                                                    ISBN 90-277-2461-X

Shang-Ching Chou: *Mechanical Geometry Theorem Proving.* 1987, 376 pp.
                                                    ISBN 90-277-2650-7

D. Przeworska-Rolewicz: *Algebraic Analysis.* 1988, 640 pp.   ISBN 90-277-2443-1

C.T.J. Dodson: *Categories, Bundles and Spacetime Topology.* 1988, 264 pp.
                                                    ISBN 90-277-2771-6

V.D. Goppa: *Geometry and Codes.* 1988, 168 pp.   ISBN 90-277-2776-7

A.A. Markov and N.M. Nagorny: *The Theory of Algorithms.* 1988, 396 pp.
                                                    ISBN 90-277-2773-2

E. Kratzel: *Lattice Points.* 1989, 322 pp.   ISBN 90-277-2733-3

A.M.W. Glass and W.Ch. Holland (eds.): *Lattice-Ordered Groups. Advances and
Techniques.* 1989, 400 pp.   ISBN 0-7923-0116-1

N.E. Hurt: *Phase Retrieval and Zero Crossings: Mathematical Methods in Image
Reconstruction.* 1989, 320 pp.   ISBN 0-7923-0210-9

Du Dingzhu and Hu Guoding (eds.): *Combinatorics, Computing and Complexity.*
1989, 248 pp.   ISBN 0-7923-0308-3

Other *Mathematics and Its Applications* titles of interest:

A.Ya. Helemskii: *The Homology of Banach and Topological Algebras.* 1989, 356 pp. ISBN 0-7923-0217-6

J. Martinez (ed.): *Ordered Algebraic Structures.* 1989, 304 pp.
ISBN 0-7923-0489-6

V.I. Varshavsky: *Self-Timed Control of Concurrent Processes. The Design of Aperiodic Logical Circuits in Computers and Discrete Systems.* 1989, 428 pp.
ISBN 0-7923-0525-6

E. Goles and S. Martinez: *Neural and Automata Networks. Dynamical Behavior and Applications.* 1990, 264 pp. ISBN 0-7923-0632-5

A. Crumeyrolle: *Orthogonal and Symplectic Clifford Algebras. Spinor Structures.* 1990, 364 pp. ISBN 0-7923-0541-8

S. Albeverio, Ph. Blanchard and D. Testard (eds.): *Stochastics, Algebra and Analysis in Classical and Quantum Dynamics.* 1990, 264 pp. ISBN 0-7923-0637-6

G. Karpilovsky: *Symmetric and G-Algebras. With Applications to Group Representations.* 1990, 384 pp. ISBN 0-7923-0761-5

J. Bosak: *Decomposition of Graphs.* 1990, 268 pp. ISBN 0-7923-0747-X

J. Adamek and V. Trnkova: *Automata and Algebras in Categories.* 1990, 488 pp.
ISBN 0-7923-0010-6

A.B. Venkov: *Spectral Theory of Automorphic Functions and Its Applications.* 1991, 280 pp. ISBN 0-7923-0487-X

M.A. Tsfasman and S.G. Vladuts: *Algebraic Geometric Codes.* 1991, 668 pp.
ISBN 0-7923-0727-5

H.J. Voss: *Cycles and Bridges in Graphs.* 1991, 288 pp. ISBN 0-7923-0899-9

V.K. Kharchenko: *Automorphisms and Derivations of Associative Rings.* 1991, 386 pp. ISBN 0-7923-1382-8

A.Yu. Olshanskii: *Geometry of Defining Relations in Groups.* 1991, 513 pp.
ISBN 0-7923-1394-1

F. Brackx and D. Constales: *Computer Algebra with LISP and REDUCE. An Introduction to Computer-Aided Pure Mathematics.* 1992, 286 pp.
ISBN 0-7923-1441-7

N.M. Korobov: *Exponential Sums and their Applications.* 1992, 210 pp.
ISBN 0-7923-1647-9

D.G. Skordev: *Computability in Combinatory Spaces. An Algebraic Generalization of Abstract First Order Computability.* 1992, 320 pp. ISBN 0-7923-1576-6

E. Goles and S. Martinez: *Statistical Physics, Automata Networks and Dynamical Systems.* 1992, 208 pp. ISBN 0-7923-1595-2

Other *Mathematics and Its Applications* titles of interest:

M.A. Frumkin: *Systolic Computations.* 1992, 320 pp.   ISBN 0-7923-1708-4

J. Alajbegovic and J. Mockor: *Approximation Theorems in Commutative Algebra.* 1992, 330 pp.   ISBN 0-7923-1948-6

I.A. Faradzev, A.A. Ivanov, M.M. Klin and A.J. Woldar: *Investigations in Algebraic Theory of Combinatorial Objects.* 1993, 516 pp.   ISBN 0-7923-1927-3

I.E. Shparlinski: *Computational and Algorithmic Problems in Finite Fields.* 1992, 266 pp.   ISBN 0-7923-2057-3

P. Feinsilver and R. Schott: *Algebraic Structures and Operator Calculus. Vol. I. Representations and Probability Theory.* 1993, 224 pp.   ISBN 0-7923-2116-2

A.G. Pinus: *Boolean Constructions in Universal Algebras.* 1993, 350 pp.
                                                            ISBN 0-7923-2117-0

V.V. Alexandrov and N.D. Gorsky: *Image Representation and Processing. A Recursive Approach.* 1993, 200 pp.   ISBN 0-7923-2136-7

L.A. Bokut' and G.P. Kukin: *Algorithmic and Combinatorial Algebra.* 1994, 384 pp.   ISBN 0-7923-2313-0

Y. Bahturin: *Basic Structures of Modern Algebra.* 1993, 419 pp.
                                                            ISBN 0-7923-2459-5

R. Krichevsky: *Universal Compression and Retrieval.* 1994, 219 pp.
                                                            ISBN 0-7923-2672-5

A. Elduque and H.C. Myung: *Mutations of Alternative Algebras.* 1994, 226 pp.
                                                            ISBN 0-7923-2735-7

E. Goles and S. Martínez (eds.): *Cellular Automata, Dynamical Systems and Neural Networks.* 1994, 189 pp.   ISBN 0-7923-2772-1

A.G. Kusraev and S.S. Kutateladze: *Nonstandard Methods of Analysis.* 1994, 444 pp.   ISBN 0-7923-2892-2

P. Feinsilver and R. Schott: *Algebraic Structures and Operator Calculus. Vol. II. Special Functions and Computer Science.* 1994, 148 pp.   ISBN 0-7923-2921-X

V.M. Kopytov and N. Ya. Medvedev: *The Theory of Lattice-Ordered Groups.* 1994, 400 pp.   ISBN 0-7923-3169-9

H. Inassaridze: *Algebraic K-Theory.* 1995, 438 pp.   ISBN 0-7923-3185-0

C. Mortensen: *Inconsistent Mathematics.* 1995, 155 pp.   ISBN 0-7923-3186-9